Rebbecca, oops

Enjoy the Book

Mind Schtin :)

Bulletproof Principles For Striking Gold

MICHAEL P. SCHLAPPI

Jacket Design by Jim Knight Design - Provo, Utah

Jacket Family Photography by Doti

Printed by Publishers Press, Salt Lake City, Utah

First Printing May 1998

Library of Congress Cataloging-in-Publication Data

Schlappi, Michael P., 1963-
    Bulletproof Principles for Striking Gold

ISBN 0-9659559-7-4

Printed in the United States of America

10 9 8 7 6 5 4 3 2 1

For Sue - My Light, My Life

ACKNOWLEDGEMENTS

So many people have played roles in the writing of this book. First, I wish to thank my parents, Larry and Patricia Schlappi, as well as my siblings, Scott, Julie, Collette, Jennifer, Todd, and Tyler for being there when I needed them most. It is these people, together with my friends and neighbors, who gave me reason for living.

Professionally, I am indebted to Marlin Shields and the entire force at Intermountain Health Care, whose support and encouragement have been both timely and helpful.

Additionally, I want to thank Philip J. Palmer, who originally shared his writing talents in the conception of this book. I likewise appreciate my editors, including Cheryl Boyle and Gladys Margetts. I also want to thank Brent Yorgason of The Lincoln Institute for his support and assistance, and for writing the foreword to this book. Finally, I wish to express appreciation to Lighthouse Publishers for believing in my story.

FOREWORD

As I have attempted to climb into the minds of others, individuals whose lives have impacted my own, I have made marvelous and revealing discoveries. Among these findings are the common core elements that combine to allow a person to climb the highest mountains life can offer. True greatness-the final measure of a man-

Seldom during a lifetime does one chance to meet-and then be influenced by-a person whose spirit permeates greatness. For me, this "chance" meeting was one beautiful April evening, in 1981. I had been asked to speak to a youth group at Mountain View High School, in Orem, Utah, and the person conducting was student body President Mike Schlappi. He was not only cheerful and charismatic, but more importantly he reached out to shake my hand and greet me with a genuine maturity and interest that was totally unexpected.

Since then, I have watched Mike's life and career from a distance, captivated by his choices, his dogged dedication, his commitment to excellence, and finally his unassuming resolve to internalize personal integrity as the foundation stone of his life. He is not only a man to be admired, but one to listen to, to learn from, and then to emulate.

BRENTON G. YORGASON

CONTENTS

1

THE DAY IT ALL BEGAN
Friday afternoon at 4:00 –
November 11, 1977

> *"A bend in the road is not the end of the road*
> *unless we fail to make the turn."*
> ANONYMOUS

BANG!!!

As the lone .38 caliber bullet tore a massive hole through my shirt and ripped into my chest, it plummeted me back against the headboard of the bed. My body convulsed violently while my mind struggled to comprehend. Unbeknownst to me, the bullet had brushed past my heart, then slammed into my back-bone, causing me to instantly lose all comprehension of feeling-

"Quit kidding around!" my friend screamed, while at the same time throwing the gun toward the corner of the room. "Quit it! You're faking it!! Get up. . .you're faking it!!"

What's happening? I asked myself. Why is he screaming? Why can't I breathe?

As these questions scanned in slow motion through my uncomprehending mind, I grabbed my shirt and pulled it up for him to see. The wound was staggering, and the blood was spreading across my chest. As my friend saw the damage, then ran from the room in shocked terror, I reached for the phone to call my mom. But. . .my legs wouldn't respond, and even more confusion ensued. Although I didn't realize it at that moment, the bullet had pierced my right lung, and breathing was becoming almost impossible.

My lung was collapsing, and strangely I could feel it filling with blood. Instantly, my entire frame seemed to be glued to the bed, waiting for my next command, but knowing it could not respond. I was becoming increasingly dizzy and lightheaded. . .and while I could hear my friend yelling and throwing dishes in the kitchen, and pounding on the organ keys in disbelief, my only thought was that he needed to call my mom-if I had any chance at all of surviving the next few minutes!

Reaching down with my right hand, I grasped my thigh. But there was no feeling. I tilted my head to my right, and looked down along my body to confirm my greatest fears. Although I could equate a tingling sensation inside my leg, there was no feeling at all from the contact my hand had made on the outside.

Time seemed almost suspended, and while it seemed like an eternity, sixty to ninety seconds later my friend reentered the room and hunched down next to me. He had a look of panic in his eyes as he grabbed my shoulders and shook me.

Retreating to the phone nearby, he instinctively dialed my number. Horror riddled with disbelief etched itself in his face as he stammered, "Mrs. Schlappi, come quick! Mike's been shot!!"

While his words floated through my puzzled mind, my initial response was Why me? I'm an athlete. . .I need my legs! I just didn't get it.

Moments passed, and while they did, I silently blacked out. Somewhere in the midst of this nightmare of increasing pain, I promised God that if I lived, I would be a better person; and that I would share my experiences with the world. How much of this I promised at that moment, I really don't know. What I do know is that my mind became increasingly fuzzy, and then the lights went out.

Although my body needed to protect itself from the assault of pain, I awakened a minute or so later upon seeing my aunt, Janet Tuttle. She was lifting my legs onto the bed, moving me into a more stable prone position. I then recognized my mother as she entered the room and threw herself toward me on the bed.

"Gee, Mom," I quipped, my mind somehow sensing the need to calm her fears, "you look nice today." Her hair was in curlers, as she was still "preparing" for the afternoon's football game.

"Calm down, Mike, and I'll call the ambulance. . . ." As her words trailed off, I instantly knew that I would be okay. I would survive, because my mom would take care of me. I then passed out again, hoping that my dream-or nightmare as it were-would soon be over.

My next recollection was being wheeled on a stretcher into the emergency room of the local hospital. The doctors were hovering over me, examining my chest; and were telling my mother that I would likely not make it through the next few hours. It was that point-blank. At that moment, I knew that something terrible had gone wrong. In one tragic instant, I lost the feeling and movement in both of my legs, and for the first time I began to realize that, barring a miracle, I'd be spending the rest of my life in a wheelchair. I lost the use of my abdominal muscles, I lost the ability to control my personal hygiene. . .but most of all, I lost my identity and my future. What's more, on a much lighter note, my football teammates would dedicate the game to me, and then promptly lose by a gut- wrenching score of 36-0!

Unaware of these details about myself, but acutely aware of the grave nature of my accident, a new game for me had just begun-only this time with a much more formidable opponent than a football team. To be victorious in this new game, I had to learn to focus on what I had remaining, rather than on what I had lost. Closing my eyes, I then made a promise to myself that I would not die on the operating table, while the doctors inserted a tube to drain the fluids from my lungs. Furthermore, I again prayed, pleading with the Lord to spare my life-and that if He did, I would spend it helping others. I realized that I had begun the physical battle of my life-and although I didn't know it at the time, the Lord would indeed answer my prayers. Even so, from that moment on, physical and psychological battles would become a daily challenge, and my ability to "do life" would be in His hands.

Several hours later, as I regained consciousness from the procedure that did, in fact, save my life, I immediately began to reflect upon the events leading up to my being shot. It had begun as a glorious November day in Orem, Utah, with the final colored leaves falling to the ground victims of a light fall breeze-

*   *   *

The screen door flew open and slapped against the side of the house. Instantly, I leaped down the three front steps and quickly cut right as if I was avoiding a tackle in the approaching championship game. My cleats dug into the freshly-cut lawn as I sped through two neighbors' yards, then across the road to my friend's house.

> *"No matter what we have done up until now, our future is spotless."*
> ANONYMOUS

Today was a great day, one I had looked forward to for months! Our team had survived the Little League All-Star football playoffs, and for the championship we were to play our cross-county rivals from Provo! Three weeks shy of my fifteenth birthday, I had been blessed with unusual coordination. Because I was the quarterback of the team, and was responsible for leading the team to victory, I was pretty serious about my role. It wasn't something I gloated about, but inwardly I felt fortunate-an athlete with promise and potential!

Arriving at my friend's home, I rapped solidly on the door. I was feeling ten-feet-tall, and my adrenaline was pumping so fast I could hardly wait to get to the game. I had never been accused of having patience, especially in waiting for a big game. Basketball, baseball, and football were my life-my very identity.

"Come in, the door's open!"

As my friend called from the back bedroom, I entered the house and headed back to where I had heard his voice. He was still getting dressed, so I just sauntered around. The grass clippings on my cleats had left a path from the front door to his bedroom, and continued to leave traces wherever I walked. But I was a fourteen-year-old boy, and as such was oblivious to the mess I was making.

Sitting finally on the edge of the bed, I looked curiously around and noticed his father's .38 caliber revolver on the night stand. This was not out of the ordinary, however, because his dad was a police officer and this was his off-duty pistol. Immediately intrigued, I leaned over and picked it up. It was in a brown leather case, and the flap was snapped closed. My only thought was how heavy the pistol was!

By this time, my friend was nearly ready to go. He came into the bedroom where I was holding his father's revolver, and smiled. "Hi, Mike! Ready to kick some Provo butt?!!"

As he spoke, he reached over and took the gun from my hands. He then unsnapped the leather cover, while I plopped down on the bed in anticipation of getting a closer look at such a magnificent weapon. Without speaking, he flipped opened the cylinder. I watched, transfixed, as the stainless steel bullets fell harmlessly to the bed. One, two, three, four. . .five. I counted the bullets as they lazily bounced onto the white bedspread. They looked cool, all silver and sleek.

My friend then pushed the cylinder closed with his hand, and held the gun up. Knowing that the gun was now empty, he was simply playing around. Innocently, he pointed the gun at my chest and pulled the trigger. That instant-that agonizing, life-altering split second-became the most serious and mentally replayed moment in each of our lives. Its repercussion was immediate, although it would be months and even years before either of us would feel its full impact. Fortunately for me, the consequences would be positive, for it would set me on a course of destiny that contained blessings and promises unfathomed.

2

NEW CHALLENGES—
LIFE GOES ON

> *"How many parents, in moments of anger, push the "kill" switch of one of their kids by telling their little boy or girl that he or she will never amount to anything? How many kids then spend a lifetime working very hard to make their parent's prophecy come true?"*
> OG MANDINO

Spending the next several weeks in the hospital, then in the rehabilitation center in Salt Lake City, gave me ample time to reflect on my life. I was born in the south-central rural Utah town of Fillmore, the second of seven children. My parents were the greatest friends-and now support system-I could have hoped for. Because of our large family, and also because of our family's values, hard work and achieving personal potential were ingrained in my mind from as early as I could remember.

Eight days following the accident, I was moved out of intensive care. Feeling somewhat better, I decided to pen my thoughts in the form of a poem. Because of its positive outlook, and since it clearly reflects my nearly fifteen-year-old mind-set, I have decided to share it as follows:

The things we've lost are the things that we love
But if we live good lives, we will regain them above.
Why things happen, God only knows
But what is planned for us, time only shows.
While we are waiting to find our spot,
We need to start working with what we have got.

We can take it easy or do the best we can,
That shows the difference between a boy and a man.
Where the path is rugged and the mountain steep,
What we learn here is ours to keep.
Where our vision stops is not the end,
There is a whole new world just around the bend.

So when there is nothing but troubles and everything's
    wrong,
Stiffen your mind and stand all the more strong.
We were sent to this earth to take our test,
So the Lord can find the good, the bad, and the best.
If we live the gospel and keep on tryin'
Before we know it, we will be in Zion.

So, when all is lost, you've still got love
From your family, your friends, and your Father above.

GRANDMA'S CHERRY SLURPEE

The night after writing this poem, I received a surprise visit
from my Grandma Schlappi. Having been moved to a different
room in the hospital, I was feeling like I might even live.
Anyway, she arrived late that night with a cherry Slurpee from
7-11. She said she was determined to give me an "unautho-
rized" treat. We visited, then she left just as a nurse came to

take my temperature. I decided I'd have some fun, so after the nurse left, I took the thermometer out of my mouth and stuck it down into the last half of my Slurpee. Moments later, when I heard the nurse returning, I hurried and put the thermometer back inside my mouth, then pretended to have dozed off.

Totally unsuspecting, she came into my room, extracted the thermometer, and read it. She then went hysterical, as she could see that I had died twenty or thirty degrees earlier! By this time, I couldn't contain myself, and I burst out laughing. I thought it was the greatest joke ever. She didn't think so, however, and my antics ended up getting me in real trouble. It was worth it, though, since this was the first time I had laughed since the accident. It was at this moment that I realized something that has sustained me many times since-although things didn't always go as I would perhaps like, still I could retain my sense of humor and keep life in perspective. That "laugh" was a real stepping-stone for me in the recovery process. From it I learned that I could still have fun, as any fifteen-year-old boy could, even though I would do it with tires instead of gym shoes.

A MAN NAMED BOB

When I was finally transferred to the rehabilitation center, in Salt Lake City, I saw a patient scooting along the hallway, face down, on a cart. I could tell that he was blind, and I asked one of the attendants about him. He told me that the man, whose name was Bob, had been paralyzed in a car accident. As if that weren't enough, he quickly acquired a defeatist attitude. Sinking in despair, he finally put a gun under his mouth and pulled the trigger. The only problem was, he didn't kill himself. Instead, he instantly lost his eyesight, leaving him more disabled than ever.

So now Bob was both blind and in a wheelchair-and with a

much better attitude than he had before he shot himself. This illustrates the point that whether you have everything going for you, or are totally disabled, the difference is measured by attitude.

> *"Tell me about a man who has won the lottery, then about a man who has broken his neck, and you still haven't told me anything about either man's happiness."*
> ANONYMOUS

FAMILY SUPPORT

My brother, Scott, who is fifteen months my senior, was a special inspiration to me during my recovery. He had been part and parcel of my earliest memories, ever since we moved to the small rural Utah town of Glenwood when I was just a year old. So living there created my earliest memories, and these memories included Scott.

My dad was the basketball coach of nearby Richfield High School, and due to his influence, we grew up with a ball in our hands. It was the most natural way for us to spend our time. We also loved to fish, and to play night games behind the nearby church. In whatever I did, the common thread of "competing" and being my very best was woven into my innermost fabric. It became me, and there was nothing I couldn't do. What's more, I seemed to develop with natural athleticism, and was always called out first to be on a team. Once chosen, it fell on my shoulders to lead out and direct whatever team I was on, and I had a mind-set where I relished the challenge.

We had a hunting dog at this time by the name of Skipper. He was an English Springer Spaniel, and when he died, we took his body over to Marshmallow Meadows-a place where we would roast marshmallows-and there we had a private graveside service for him. It was sad, and we were all crying, but I appreciated my dad's sensitivity to our needs, and my own sadness. Dad was right there getting us Skipper Number Two, however, so this helped get over the loss of such a great companion.

My favorite recollection of this time, however, was of going to the high school gym with Scott and my dad, and playing basketball. We were in seventh heaven when we did this, and nothing could be greater than a good game of basketball. I was a gym rat from day one, and because I had a competitive bulldog nature, there wasn't anything I couldn't do with a ball in my hands.

When I was about six years old, Dad decided he'd better get into a profession that was more financially lucrative, so we moved to Orem, where he became an insurance agent. Some insurance agents might have said that coaching was more lucrative, but he wouldn't have listened! Anyway, the change in professions worked, and he later founded his own financial planning firm. He always worked hard, providing me with the greatest of role models.

> *"Nothing is really work unless you would rather be*
> *doing something else."*
> SIR JAMES BARRIE

CHORES AFTER MY ACCIDENT

When I returned home from the hospital following the shooting, I was prepared for some serious sympathy. What I didn't know, of course, was that my parents had discussed this situation at length, and had determined to treat me no differently than they had before.

Dad was more a pusher than Mom, and she was more the

loving, compassionate mother with unconditional love. Both parents were supportive, of course, but Mom was especially helpful in encouraging me to regain my self-worth, an identity that, previous to my being shot, had revolved around the strength and coordination of my hands and legs.

> *"If you have a big enough WHY in life, you can always discover the HOWS!"*
> ANONYMOUS

My jobs were still the same, and with the garden this meant that I had to wheel myself along between the rows of vegetables, weeding by hand with no excuses. I wasn't exactly excited about this at the time, but their wisdom caused me to eliminate any potential for "pity parties," and I moved ahead thinking of myself as normal-no disability.

Another chore that remains vivid in my memory was the daily routine of setting the table before a meal, then washing the dishes afterward. I would stack the used dishes in my lap, take them from the dining room to the kitchen, then sit there sideways and wash or rinse them off before putting them in the dishwasher.

My folks were smart enough to let me figure out the "hows" for myself. They supplied the "what" and there was really no discussion afterward. And so, with their approach to my situation, I soon learned to cope with my responsibilities-to do things like my siblings did them even though in a wheelchair. Looking back, in the long run I was much the better person for it.

I don't mean to infer that my folks didn't find things for me to do that were more suitable to my condition. In fact, whether it was fixing a sprinkler head that had broken, or vacuuming the part of the house where I lived, there was plenty for me to do. Being limited in what I could do, I was forced to think ahead, to anticipate my needs. I'm sure this forward-looking way of doing things has spilled into all areas of my life, rather than remain a responder.

> *"Great things are accomplished by those who know how, but they will always be led by those who know why."*
> ANONYMOUS

One of my greatest blessings at this time was my sister, Julie. She was three years younger than me, but even so, she grew up in a hurry following my accident, and quite literally became my own private nurse. There was nothing she wouldn't do for me, and even though we had very normal brother/sister squabbles, we also grew very close.

My other siblings helped, as well, and with their support and encouragement, I soon found myself rekindling the same attitude of success that had sustained me thus far in my life. Deep inside, I knew the power of my mind, and within limitations, there wasn't anything I couldn't do. For me, it was the perfect place to begin my long road back.

3

RACING AHEAD–
BEGINNING ATTITUDE
THERAPY

> *"So often we seek for a change in the situation*
> *when all we have to do is change our attitude."*
> ANONYMOUS

ATTITUDE THERAPY DEFINED

I received a lot of physical and occupational therapy following
my accident, and I was plenty grateful for it, believe me. Even
so, my greatest therapy was the therapy I gave to myself. I
called it Attitude Therapy. This notion evolved, and is now the
name of my speaking business. I've often heard it said that "our
attitude determines our altitude," and I believe it. Regardless of
how much help I received from those attending me, I knew
intuitively that the greatest therapy was in my mind-and consisted of what I convinced myself to do and to become. I knew
that I would have anatomical limitations, to be sure, but I also
knew that within these limitations I could achieve anything I
determined to achieve. From where I sat, there was no holding
back, and I began to relish the challenge of each new day!

Charles Swindoll, in speaking about attitude, said something
that deserves restating:

"The longer I live, the more I realize the impact one's attitude
has on life. Attitude, to me, is more important than facts. It is

more important than the past, than education, than money, than circumstance, than failures, than successes, than what other people think or say or do. It is more important than appearance, giftedness or skill. What's more, it will make or break a company, a church, a home-and of course, a person.

"The remarkable thing is we have a choice every day regarding the attitude we will embrace for that day. We cannot change our past, we cannot change the fact that people will act in a certain way. We cannot change the inevitable. The only thing we can do is play on the one string we have, and that is our attitude.

"I am convinced that life is ten percent what happens to me, and ninety percent how I react to it. And so it is with you. We are in charge of our attitudes!"

Adding to this, Dallin H. Oaks, a noted Supreme Court judge, once said that three ingredients comprise the inner man. These include (1) our motives, (2) our desires, and (3) our attitude[1]. In his preface, Mr. Oaks states that "Motives explain actions completed. Desires identify actions contemplated. Attitudes are the thought processes by which we evaluate our actions and experiences. Motives, desires, and attitudes are interdependent."

With these thoughts in mind, I would now like to share an experience that happened to me about a year after my accident.

DEER HUNTING-WITH A NEW PERSPECTIVE

When I turned sixteen, one rite of passage that seemed pretty important to me was to purchase my first deer hunting license, then go hunting with my dad and brother. I'd legally hunted pheasants since I was twelve, so the yearly fall ritual was deep

---

1. Pure In Heart, Salt Lake City: Bookcraft, 1988.

inside me. Even though I would be hunting deer from a sitting position against a tree, and would not be able to stomp through the trees and brush as before, the effort would be well worth it.

The opening morning of deer hunting season finally arrived, and before I knew it my arms and legs were wrapped around Dad on his horse, and we were moving up the slopes in the mountain range east of Fillmore, Utah. When we finally decided on a good spot for me to position myself, I was lifted off the horse, propped up against a large pine, and handed my rifle. Dad and the others then moved on with the objective of working below me and hopefully moving a deer up through the trees for me to shoot.

I had not been there long when a herd of thirteen or fourteen elk came along, not even noticing me. I watched them, totally enthralled with their beauty and majesty. Actually, if I were to be totally truthful, I was scared to death. I could just see them trampling on me with no way to save myself. After they left, I heard other noises, probably deer, over the ridge; but in my mind I could just see a bear coming toward me. My mind really began playing tricks, and I had a hard time settling down.

Not long after this, a nice two-point buck came into the clearing below. I was ready, too, and I swung my rifle to my shoulder, pulled the trigger, and dropped it with one shot. It was a nice first time trophy, and I couldn't have been more excited! After years of hunting with my dad, and carrying our lunch, I had finally bagged my very own deer!

Not long after this, another hunter came along. What I hadn't known until then was that, tragically, the deer hadn't instantly died when I shot it. I had only broken its back. When the other hunter came along, the deer jumped up on its front legs and began dragging itself along with a broken back, trying to escape. The hunter shot it again, and this time it stayed down. It was just about fifty yards from me.

I yelled, "Hey, Dude!" and the man came over to me on his horse. As calmly as possible, I explained that I had already shot the deer. Seeing my plight, he was pretty cool about it, and asked if he could gut it for me. He could see that I was paralyzed and not going anywhere, so he really went out of his way to help. I thanked him, and before I knew it he had dragged the deer over to a nearby tree, cleaned and dressed it, and left it hanging to cool off. I thanked him sincerely, and he was once again on his way.

Later in the day, when my dad and the others returned, several of them had likewise bagged their bucks. Were they ever surprised, though, to see that not only had I killed my first deer, but it was dressed and ready to carry out. They wasted no time, and soon the deer and I were hoisted up onto the horse, and we were headed back to camp.

Earlier in the day, when I saw that my deer hadn't initially died, I suddenly had a surge of feelings of empathy and sadness that I had never before known. In actuality, the deer was no different than myself. Because I had pranced a mile in its hoof-prints, I knew what it was experiencing-the pain, the sense of hopelessness, the broken back, the loss. What's more, I didn't have a wheelchair to offer the deer, nor a pain-killer to lessen its pain.

Without intending to be over dramatic about this moment, my heart suddenly felt differently about killing than ever before. I knew the Lord had provided for our family to have an increase of meat for the winter, and I'd had a great time; but I somehow valued life, and the deer's loss of agency, in an entirely different way.

> *"The same refinement which brings us new*
> *pleasures, exposes us to new pains."*
> RICHTER

A NEW PERSPECTIVE ACCOMPANYING MY ATTITUDE

I'm not sure of all that happened to me on that deer hunt, but from this time forth I began to take life a little more seriously. In fact, since then I've never killed a deer, but to this point have been satisfied in shooting them with my camera.

Not long after this, I went with my family and my friend Roger Dayton to Deer Creek Reservoir to do some waterskiing. We had a great day, although I was pretty much confined to the boat. I hadn't yet learned to water-ski, but I had learned to have fun.

After all the others had finished skiing, it was suddenly my turn. So I put on a life jacket and flipped over the edge of the boat into the water. Perhaps I should mention that at this time there were no special water skis for the disabled. All we had was a round yellow plywood board that my friend, Roger, had brought along. So I pulled myself onto this board, grabbed hold of the rope, and signaled for my dad to open up the throttle.

He took off at full speed, and I hung on for dear life. I can't say that I wasn't frightened, because I was. Still, even though the water was hitting me in the face and making it hard to see, I was doing just fine! It was then that I noticed the people in our boat laughing. I thought that perhaps one of my sisters, Julie or Collette, had told a joke; but I remembered that neither of them were that humorous. Then, almost immediately Dad swung the boat around and headed for shore. By this time, I was becoming self-conscious, and thought that perhaps my legs were flapping around in the breeze, giving everyone a good

laugh. I thought that perhaps I looked like an injured
Superman, or like the flying frog in the Budweiser beer com-
mercial. I was just skipping along the water, having a good
time!

At this moment I glanced behind me, and was horrified to
learn that the water had pushed my swimsuit completely off
my atrophied legs! I was literally skiing in my birthday suit!
Well, my brain wasn't paralyzed, so I quickly slid off the board
and let go of the rope. I then waited for Dad to circle around in
the boat and pick me up.

I was mortified, to be sure, but while I bobbed up and down
in the water, I suddenly realized that this moment was much
like my accident. I couldn't help what had happened to me, but
my attitude could determine how I responded to it. I could
either laugh at myself, and enjoy the humor of the moment, or
I could sulk and get upset with how the others were acting. It
was a great "wake up" call for me, and a memory that had
much more to do with my future than did my lost swimming
trunks!

Somehow I regained my dignity, borrowed an extra pair of
shorts from my sister, then got back in the boat. We had a good
time laughing and joking as I told everyone I was the world's
first nude disabled water-skier. Let me add something profound. I experienced firsthand that having a "bad butt day" is
worse than having a "bad hair day!"

On a subsequent trip to the same reservoir, I had an even
greater tragedy-but with very severe consequences. We were driving down the canyon after waterskiing, with Roger Dayton in
his very antiquated truck. Although I didn't know it at the
time, I was about to have another extreme learning "moment".
Although I was unaware of it, the truck floorboard had a predictable habit of becoming very, very hot. With no feeling in

my feet, and with nothing covering them for protection, I was oblivious to any problem until I began smelling barbecued meat. What I didn't know was that I was the barbecue, with my feet being burned all the way through to the bone! The only good thing about this was that I had no pain. The bad thing was that I wasn't able to wear shoes for about six months after that!

The point of sharing this story is that the perils and consequences of being paralyzed are many, and often unpredictable and unanticipated. I now had to be more aware than ever before-either that, or I would spend my life banged or burned up!

Even more difficult than pushing a wheelchair around were these constant threats to my safety. I also had a hard time dealing with the phantom pains in my legs, the bathroom complexities, and so forth. I soon learned that although this was reality, I was no different than anyone else. Just because others don't have a wheelchair strapped to their back, that doesn't mean they don't have their challenges. Life is a challenge, and must be faced with courage and resolve.

Still, even with an overall positive attitude, I had my moments of near discouragement and despair. One of these moments was not long after I got out of the hospital. I was back in school, and one afternoon my buddies and I went to the sophomore basketball game. Two friends carried me up into the bleachers, and as the game unfolded, I could hardly watch. Instead, I quietly wept, feeling sorry for myself and despairing over the fact that I would never again play basketball on my feet with my brother, Scott. It was a dark hour for me, one that I still remember as though it were yesterday.

About two years after the accident, my mom bought me a record by Barry Manilow. The name of one of the songs on this album was, I Made It Through the Rain. It was a beautiful song, and since she thought I had progressed through the worst of things, she sensed it would be a comfort to me. It was, too, and I appreciated her sensitivity. I was on an unexpected learning curve with one storm after another, but I was amazed at how far I had progressed in three short years.

> *"The best predictor of future behavior is past behavior."*
> ANONYMOUS

PARADIGMS OF PURPOSE

It has been said that the best predictor of future behavior is past behavior. I knew that if I was going to get back to my productive, success-oriented self, I would need to grow up in a hurry and face life straight on. This meant that I would have to think in a much more mature manner, and that I could not return to my adolescent carefree mentality.

Don Hutson, in his book The Sale, stressed the point that looking back is a useless emotion with no intrinsic value. If we could put ourselves (or even others) into some kind of rewind mode, or if we could relive an event, it might be useful. But we can't do this, so we shouldn't burn up energy looking backward. Instead, we should learn from the past, then look forward with "an eye of hope."

Returning to my observable improvement, as I continued to take stock of my life, I came to several life-altering shifts in my mind-set. That is, I began to view life in a different way-seeing things from a different perspective, and with a much clearer lens and focus. These life views are called paradigms, or expectations. They followed what could have been a tragedy for me, and have built upon each other in these sequential steps:

Paradigm One: Small successes, if repeated, lead to large successes. For me, this began with simply being able to partially dress myself. Now, years later, caring for my personal needs is relatively easy and automatic. While I have to often become creative in how I accomplish things, there is nothing I can't do. It's a great way to live life. Confidence is born of small successes, and having an increase in ability makes it all worthwhile in the end.

Paradigm Two: Passions don't die simply because we are injured, physically or emotionally. For me, this meant that I decided to pursue life with the same enthusiasm I had prior to my accident. After all, my spirit wasn't paralyzed, and I could still soar with the other eagles! Akin to this idea is that we should focus on what we have left, rather than on what we may have lost. Having this mind-set has helped me want not to quit, want not to become a disabled victim, and want not to ultimately become a burden to society.

> *"All motivation is self-motivation."*
> ANONYMOUS

One of the most inspirational examples of this concept was recently reported in The Daily Herald[2]. In Ed Heroldsen's article, he spoke of a senior citizen named Thora DeWolf, who drove alone to her second home in Nauvoo, Illinois, just last April. Four days after her arrival, she fell and broke both shoulders which left her helpless and alone for five long days. Unable to eat or drink, and to even move, this courageous woman maintained hope that she wouldn't die there. On the fifth day,

---

2. Saturday, February 21, 1998, C3

Thora was able to turn over on her back. This allowed her to call out loud enough to be heard by a neighbor taking a stroll.

Thora was rescued, then flown to a hospital in nearby Keokuk, Iowa. After several weeks in the hospital-where she was not only bedridden, but was unable to feed and care for herself-she gradually began physical therapy. This consisted of moving her hands in small circles several times a day. This ability led to another, and before long she was not only out of the rehab center, but was doing community service on a full-time basis.

Reflecting on her recovery experience, Thora states, "When I first started to exercise it was painful and I was overcome at the end of the day. But the next morning I would feel so wonderful! I got the message that if I ever wanted to be normal again, I had to do it because no one else could do this for me."

Talk about one who is in charge of her life! Perhaps the most remarkable aspect of Thora's experience is her age. She is eighty-two years young, and still awaking each day to serve-living her passion in every way!

Paradigm Three: Regardless of our limitations, we don't need to consider ourselves "handicapped." For me, this meant that I teach myself to not use my wheelchair as a *crutch!* This is not to say that I didn't have to be on guard, for just the opposite was true. Still, even though I could barbeque my feet, or lose my shorts in a very unique manner, I was no different from anyone else. I had my challenges, but so did everyone.

We all have challenges, whether they're obvious like mine, or more hidden, such as the emotional trauma of a death or divorce. In a recent lecture given in Salt Lake City, noted psychologist Dr. Page Bailey taught that injury is an interruption of expectations. Everyone-at some time or another-experiences injury. It is then up to that person to create new expectations, and to not allow their misfortune to derail their expectations and leave them in pain and anguish.

Regardless of my injury, I determined to never consider myself handicapped. My expectations had been interrupted, but only for a short time. I soon learned to expect as much from myself as before the accident-maybe even more. This is why I have used the word "disabled" throughout this book, rather than the debilitative word handicapped.

Paradigm Four: We should not blame others when "bad" things happen. In other words, we should be responsible for ourselves. I first learned this lesson when I was about seven years old. One day, while visiting my folks' bedroom, I noticed a red cherry-looking ball on the top of their dresser. It had what looked like a fuse on it, so I stuffed it into my pocket and left the room. Devising a very complex plan, I found some matches, retreated to the back yard under a willow tree, and set the fuse on fire.

Jumping back, I was fortunate not to be injured when the "cherry bomb" blew! The entire neighborhood heard the noise, and my goose was cooked in one inescapable instant. Not only that, but the bomb had blown a foot-wide hole in the grass, and when my folks learned what I'd done, I was in the doghouse for sure! I had to take responsibility for what I'd done, and so I did, although for the life of me I can't recall the punishment I received.

By the same token, when I reflected on my shooting accident, I honestly didn't blame my friend for what happened to me. I was in his parents' bedroom, I was handling his dad's pistol, and just because he pulled the trigger and accidentally shot me, that didn't make him solely responsible for what happened. We were in it together, and while I became the initial victim, the subsequent blessings in my life far outweigh the limitations. In fact, because of this accident, I may have even accomplished more than I would have if I had been able to continue employing the use of my legs!

ATTITUDE THERAPY SUMMARIZED

As I have attempted to articulate, Attitude Therapy is the therapy we give to ourselves. It is not occupational or physical therapy; nor is it speech therapy or recreational therapy. Rather, it is seeking to recover "from the inside out." This is the magic of attitude therapy. If we can incorporate this self-therapy without receiving our doctor's degree, then it behooves us to consider it. After all, from time to time each of us needs to sculpt, change, or re-frame the way we think.

GIRLS, GIRLS, AND MORE GIRLS!

Shifting gears, part of my maturing response to life [and therefore my attitude] included feelings for and experiences with the opposite sex. While I wasn't the greatest catch in our high school, somehow I didn't know that, and after turning sixteen, I began to date with great regularity. I not only had an electrifying handshake, but I also did wheelies in my chair, and in general "did my thing" with the young ladies. Now, to be truthful, I haven't met very many girls who made it a practice to follow handicapped license plates home, and who preferred dating wheelchair-bound guys. Even so, I decided early on that I would not accept pity, nor would I view myself as being disabled, per se. Rather, I was just a guy who had to dance in my wheelchair.

This realization didn't come easy. In fact, I really hesitated to ask a girl to dance with someone who couldn't dance. Then something fortuitous took place. I went to one particular dance, and before long a girl by the name of Wendy invited me onto the floor. I was pretty slow in rolling on out, but before long I was doing the wheelies I mentioned earlier, twisting around, and goofing off until the song ended. Thanking her for asking me, I was about to turn and head back to my safe corner. After all, I didn't even want to be at the dance, let alone on the dance floor.

Wendy then caught me completely off-guard by inviting me to dance the next song-a slow one, no less! I was stuck, and I knew it, so waiting to see what her next move would be, I just sat there-as if I had any choice! Neither of us knew what to do at that point, so she approached me, then got down on her knees next to my chair. Now the pressure was on me, and [pardon the pun] I was sweating bullets! Not knowing I had better options, I starting patting her on the head to the beat of the music. Before long, however, I invited her to sit on my lap so we could coast around in each other's arms. My mom hadn't raised any dummies, I assure you, and suddenly I was floating on clouds, realizing yet another advantage to being wheelchair bound!

That initial slow dance with Wendy led to others, on other evenings, and gave me a chance to have my dates sit on my lap both while dancing a slow dance, and at picture-taking time. Their perfume was intoxicating, and keeping socially active was as therapeutic as any other aspect of my teen life. What's more, as my wheelchair antics became patented, my buddies would take my extra wheelchairs to dances, just to have their dates sit in their laps. Those wheelchairs really were in vogue, and provided us all with great moments of closeness with the girls!

> *"Let challenges make you a better person, and not a bitter person."*
> ANONYMOUS

CAUGHT WITHOUT A CHAIR

Sometime later, as I accelerated my activities as a normal teenager, Halloween rolled around. Not one to stay home and hand out treats, the night arrived with several buddies and I chasing around in one of their cars, seeing what evil deeds we could accomplish. I was sitting in the middle of the back seat, flanked by two friends. We were chucking tomatoes and water balloons at the trick-or-treaters, scaring them, and in general having a lot of fun.

Unfortunately, the local police caught on to our pranks, and pulled us over. They ordered all of us out of the car, and my buddies quickly responded. But, as you'd guess, I could do nothing but sit there. When one of the cops challenged me, I told him I couldn't get out because I was in a wheelchair.

"Where's your chair?" he challenged, thinking he had me dead-to-rights in a lie.

"Uh. . .I left it home."

"Sure you did. And tonight's not Halloween, either!"

Thinking I was being a smart alec, the policeman then grabbed me by the shoulder and pulled me out of the car. I complied, of course, but with disastrous consequences. As I came out the back door, I fell hard against the road. Shocked into the reality of my truthfully being disabled, the cop helped me back up into the car, totally tongue-tied. I'm sure that's an experience he'll never forget. I won't either, since I had to do several hours of community service!

Reflecting on these early experiences, I have thought of how our society has labeled people-whether they have too many zits, can't dance slow dances, carry an extra fifty pounds, don't wear designer clothes, or even get confronted by a cop who thinks we are someone we're not on Halloween night. We've been conditioned to be picky and judgmental about non-relevant issues, creating stereotypes that really limit our vision about people. I've found that the happiest, most enjoyable people to be around are those who look for the good in others, then are blind to whatever "deficiencies" they might have. After all, we all have deficiencies, and that's called LIFE!

4

REACHING OUT—
THE MAGIC OF SERVING

"If your lot in life feels empty, build a service station."
ANONYMOUS

Before speaking of how I have grown through serving-which, by the way, I now believe is the main ingredient to having a happy life-I would like to examine this orientation from another perspective. Following my accident, I struggled with accepting help from anyone. I just wanted to "do it myself." What I soon learned, however, is that when I allowed someone else to help me, and serve my needs, they were growing and feeling better about themselves than ever before. It's much easier to give than to receive-at least for me. The idea of letting others do things for me did not come easy. Gradually, however, I became aware of how crucial gracious receiving is-both for the giver as well as for the receiver. It helps everyone!

And now the other side of the coin.

Although I try to keep my weaknesses pretty much to myself, I haven't always made a conscious effort to think of others. In fact, when I was a little boy, I thought the world sort of revolved around me. Not only was I gifted athletically, but I was also quite quick, mentally. I always prided myself in learning my times tables ahead of the class, impressing my teachers, and so forth.

One time, however, when I was in the third grade, I had an experience that jarred me out of my self-serving moorings. I misbehaved in class-although I don't really remember what I did wrong-and my teacher, Mrs. Warner, grabbed me by the ear and escorted me to the principal's office. It was a sobering moment, for sure, and a dose of humility that I must have needed at that juncture of my life.

One of the ideas that came to me early, probably when I was heading into the sixth grade at Westmore Elementary School, was that I wanted to be student body president of every school I attended. This dream became a reality, as I was elected president of my elementary school my final year there. That experience did as much for me as athletics did, and when I entered junior high and high school, I was elected to the same office.

The other experience that I had early in my life that caused me to reach beyond myself was getting involved in Cub Scouts as well as the Boy Scouts, I earned the two highest honors in these programs-Arrow of Light, in Cub Scouts, and the rank of Eagle, in Scouts. I never once thought that I couldn't achieve these ranks, and in doing so I began to consider the needs and feelings of others.

> *"You can make more friends in two weeks by becoming interested in other people than you can in two years by trying to get other people interested in you."*
> ANONYMOUS

Both of these scouting programs were service-oriented, and I especially appreciated my Eagle Service Project. For this, I organized a program that would enable elementary-age children to

safely walk the mile from school to the local church youth
meetings on Wednesday afternoons. A lot of construction was
going on in the area, so their safety was a crucial problem. I
organized and enlisted adult crossing guard volunteers, as well
as a system that ensured safe passage for these kids. I supervised
this program for the two months of construction on the roads,
and when it was over, I felt like I had really helped out.

Still, with all this service mentality, following my accident I
honestly thought that I had lost the ability to serve. I no longer
had use of my legs, and somehow I equated legs with service. I
would see my buddies helping move a family in or out of their
home, and I would want to help, too. But I couldn't. Instead, I
would stay home, feeling limited in my capacity to give.

OATMEAL CEREAL

One morning, several years later when I was living in
California, I decided to cook oatmeal cereal for breakfast. I had
the water boiling, and was pouring the flakes into the pan
when I unknowingly spilled some of the oatmeal flakes onto
my lap. I was all spiffied up in a suit and tie, ready for a successful day. Immediately, I went out the front door to brush the
flakes off my lap. Sitting in my wheelchair, I let my front
wheels down off the curb to make it easier for me to brush the
flakes off my lap and into the gutter.

Tragically, I lost my balance and tumbled out of the chair and
into the gutter. My wheelchair also tipped over next to me, and
I looked terrible and felt even worse. My next thought was,
Well, I'll just pull myself upright, straighten my chair, then
climb back into it hoping no one would notice my faux pas. As
luck would have it, however, at that very instant a black Trans
Am sports car rounded the corner and headed up the street
toward me. Hoping they wouldn't see me, I tried to hide
behind my wheelchair in the gutter. This ploy didn't work,

however, as they not only noticed me, but stopped and rolled down the window to offer their assistance.

A young lady on the passenger side, upon noticing me amongst a pile of oatmeal flakes, called out the window, and asked, "Sir, may we help you?"

"No, thanks," I quipped, trying to be funny. "I'm just having breakfast."

Feeling uncomfortable, the girl rolled up her window, and they sped away.

As I made my way back into my chair and into the house, I felt badly for two reasons: First, I had not allowed someone the opportunity to serve me when they wanted to; and second, they may never again offer their assistance to a person in need. I had really blown it!

Since then, whenever someone asks me how they can help someone who is disabled, I instruct them to do the very thing this young lady did-to ask-without evaluating a person's response. This not only opens the door for assistance, but allows the recipient to say "yes," or "no," depending on their own needs and perspective. I believe courtesy is always in fashion, whether toward someone who is disabled, or simply in need. If the recipient does not appreciate your courtesy, then perhaps the problem lies with them. Still, you can feel good about offering. Rather than ignoring an opportunity to serve, I always say, "When in doubt, help people out."

In an incident similar to this, I was leaving a shopping mall one day when I noticed a lady next to me in a business suit, who was walking toward the same exit. When we were both about thirty feet from the door, my thoughts were whether or not I should open the door for this business woman. I'm sure she had the same thoughts about helping me. Without waiting for her to speak, I looked at her and said, "I'll get the door for you this time, and you can get it for me next time." She thanked me, and we were both comfortable in leaving the mall.

From this incident, as well as many others, my thinking gradually evolved, and I realized I could serve every bit as well from a wheelchair as I could from a standing position. I would simply focus on my strengths, and help people however and whenever I could. In fact, my favorite type of service-that which produces the most profound sense of fulfillment in my heart-is visiting newly injured individuals at hospitals and rehab centers. I am able to give them hope as well as an example of what is possible to achieve if they don't give up. Finally [this time with a pun intended] I am a roll model! Because of my condition, the disabled are able to relate with me in a more natural way than they can relate to those attending them. There is instant rapport, and empathy going both ways, and each of us is made better by the exchange.

I'm sure this process of helping people re-frame their condition gave momentum for me to begin public speaking. Even today, when I need to spend more time with my family, it is difficult for me to turn down an opportunity to speak-especially to a group of young people. They need all the help they can get, so I try to meet their needs whenever possible.

> *"Do all the good you can*
> *By all the means you can*
> *In all the ways you can*
> *At all the times you can*
> *In all the places you can*
> *To everyone you can."*
> ANONYMOUS

MY "HAVENLY" HOME
When I think of the home I grew up in, in Orem, Utah, I

think of the house, itself, as well as my family. Both were all a guy could ask for. Our home was located on a half acre of beautiful orchard land, with a massive garden. I'll have to admit that I liked the orchards better than the garden, simply because of the weeding and tending the garden required. I had my very own row to hoe, and it was like a rain cloud that hung over me all the time. There were always more weeds to pull, and because I was the kind of kid to get my jobs done before playing, I never really got hassled. Of course, that's my memory, and I'll stick to it.

But the orchards, well. . .they were the stuff life was made of for a growing boy. I owned a pellet gun and a BB gun, and I would wander through the orchards, shooting birds and fashioning myself into an out-and-out sharpshooter. I would envision the pheasant hunts, and using my shotgun, and these fantasies put me in a world of its own. One time, when a bird flew overhead, I cocked my BB gun, aimed, and fired. Believe it or not, I shot the bird while it was in flight. I didn't think about any moral issues in killing birds, since it was legal to do so; but I sure did consider myself an expert marksman.

Our home was a split-level type, and with such a large family it was not under utilized by any means. My brother, Scott, and I shared a bedroom and bathroom, and even a queen-size bed. We'd get upset with each other at night whenever we crossed over our imaginary "side" of the bed, but we didn't get angry enough to stop being friends.

MY FIRST REAL JOBS
Because I wanted money to do things with, and since mom and dad wanted us kids to learn to take responsibility, we contracted to manage a paper route. Not only did I have my own route to deliver each morning, but we had to get up at 3:30 a.m. and deliver the bundles of papers to the other carriers for their

routes. I can honestly say that I hated doing this-although I for sure didn't hate getting the monthly check in the mail.

The other early-on job I had was selling donuts. I would make a few cents a dozen, but I never did have much money from it. Instead, I'd earn some money, then turn around and buy a dozen for myself and my friends. My profits always seemed to be "eaten up" with devouring donuts!

These early habits of working were profitable in the long run, as I would save a percentage of what I earned. This money went into an account for college, and came in quite handy when I finally reached that age of tuition needs.

I also got a job at a nearby restaurant. I didn't wait on tables, of course, but kept the salad bar. I would re-stock it, clean it up, and in general hover over it so that the restaurant's patrons would think it the "best in town." I enjoyed working there, too, and feeling that there was something productive I could still do. It was great therapy for me.

It was while working there, and experiencing back pain, that I made the decision to spend my life supporting my family through the use of my brain, rather than through my upper torso brawn. Adding fuel to this decision was another job I had, and that was teaching first and second graders how to read. They would be assigned to come to our house for lessons, and I would help them develop their reading skills. Looking back, I'm sure this early exposure to instructing helped me every bit as much as it helped them. From what I've experienced, teaching is the most rewarding profession there is.

Still, I don't want to paint a picture of all work, for it was not that way. My brother and I would always be building a fort out in the orchard; and we would sit there and tell stories by the hour. We'd also sneak food out of the house-especially graham crackers or cookies, and stock the fort with whatever food we could get. We'd also take off for the sandpits and hunt lizards, catching them whenever we could. Two brothers couldn't have been more caught up with just being boys.

> *"What an enormous magnifier is tradition!"*
> CARLYLE

FAMILY TRADITIONS

The other definition of my "home" was, of course, my family. With seven children to care for, there were always plenty of chores for each of us to do. There were also lots of traditions, since we were anything but sedentary. Our day would begin with an early morning devotional and family prayer, and with a spirit of unity that lasts to this day. We owned a cabin near Park City, in Midway, and often went there on weekends. Our favorite activities were snowmobiling and sledding in the winter, and motorcycle riding, fishing, playing tennis and golf in the summer. At night, we played cards and watched videos, drank hot chocolate, and in general kept ourselves going for all hours into the night.

We also had family excursions in the summertime, the most memorable being waterskiing trips to Lake Powell. One time, a year or so before I was shot, I became an expert diver at Lake Powell, doing front flips off cliffs that were thirty to forty feet high. In reality, I could have easily broken my neck or back at that time. I now work with young accident victims, in rehabilitation, and I am amazed that I survived those dives.

Holidays were especially exciting times for our family. Christmas and Thanksgiving topped the list. We lived for these celebrations, and we were always creating new "traditions." It's ironic that I was given a shotgun for Christmas at age twelve. My siblings and I were also given a Ping-Pong table, and we played a lot of table tennis. Then, after the accident, I became even more serious about it. Although I've never played table tennis in competition, I have gotten pretty good at it, even from my wheelchair.

One of the greatest legacies my parents gave me as a boy was a sheer love for being in the mountains. When people ask me what I would do for one day, if given legs to walk on, my answer is always the same-I tell them that I would spend the time running and hiking and camping in the mountains. Of course my wife would have other plans, but that's what I'd like to do! I gain almost a mystical strength whenever I spend time in the mountains, especially in the fall and in the spring. Each range is God's handiwork, and being away from the pressures of life, while enjoying their majesty and beauty, is one of my greatest pleasures.

5

CAPTURING MY HEART—
A LADY NAMED SUE
Halloween Night, 1983

MY TREAT AND HER TRICK

It was a Halloween night full of promise and prospects. Just before dark my friend Doug Jensen and I set out to trick or treat at the girls' dorms at Brigham Young University. We were both going to school, and having the time of our lives dating and socializing. On this particular night, as we were going from dorm to dorm, we knocked on one door that would quite literally become my door of destiny.

When the door opened, there stood one of the most beautiful persons I had ever seen. Introducing ourselves, this blonde girl named Sue invited us in. The boy she was dating had been unable to see her that evening, so she was home with her roommates, answering the door for guys like us who knew where the action was.

When we entered Sue's dorm, she thought I was all decked out in my wheelchair as part of my Halloween costume. I guess I really faked her out, and when I wheeled into the living room, I immediately lifted myself out of my wheelchair and plunked myself down on the sofa. I had done this hundreds of times before, so it was no big deal. But I think it was then that she realized I really was wheelchair bound, although she didn't say anything about it.

Before long, Sue and her roommates were feeding us Captain Crunch cereal and a handful of grapes, and we were laughing and having a great time. Although I was twenty-one years old, having just completed a full-time mission for my church, I was just a freshman on campus. When I met Sue, she seemed quite a bit older. But we hung out there in the dorm, and although I didn't know it at the time, Sue's roommates were surprised that she was spending time with me-especially since she had a serious boyfriend. That didn't seem to bother her, though, as we hit it off pretty well.

After a while, Doug and I decided it was time to leave, so we resumed our trick or treating. The following morning, although I was unaware of it, Sue was having a conversation with about fifteen other girls in the dorm, and they were all talking about "the guy in the wheelchair." We had visited them all, and evidently made a pretty good impression.

I purposely played "hard to get" for the next two weeks, but after determining to give Sue a call, I accidentally bumped into her as she was coming out of the university library. Well, I didn't literally bump into her, but I did see her and visited briefly with her at that time. I also got up the nerve to ask for her phone number, and although my memory may be playing tricks on me, I recall that she actually gave it to me! She said something like, "Why don't you wheel on over?" To be truthful, she didn't say that; but she did give me her number, so I called and asked her out.

Our first date was to a pre-season BYU basketball exhibition game, in nearby Heber City. This was about twenty-five miles up the canyon from Provo, and since I was the team manager, I knew it would be a good way to impress her. I was driving a 1980 two-door Pontiac J-2000 at the time, and after climbing

into the driver's seat, I'd collapse my wheelchair and throw it
into the seat behind me. I felt pretty heroic doing this, and Sue
was impressed-at least she appeared to be.

We had a good time, and visited nonstop all the way up the
canyon. Because I drove my car with hand controls, I wasn't
able to hold her hand. Most girls probably wish all vehicles had
hand controls, and I suspect she felt the same way. She was safe,
her hand was safe, and I just hoped she wouldn't lock her heart
to having a second date.

That was about it, though, since by the time the game started, I became so nauseous I could hardly hold my head up. But
Sue didn't seem to mind, and somehow I made it back to Provo
and dropped her off. To be truthful, I don't know if I was more
relieved the date was over, or if she was. I couldn't imagine that
she was impressed, especially since she thought I was sick
because I was in a wheelchair. But that had nothing to do with
it. I was just plain sick to my stomach.

Still, something good must have happened on our first date
since not long afterward we attended a play on campus titled,
"Singing Sergeants." My family met us there, so I was able to
introduce them to Sue at that time. My sister-in-law, Becky,
told me later that all she could think of was, "Oh, no. . .anoth-
er ditzy blonde!" I must have dated several of these girls,
although none of them were ditzy in my mind. I dated plenty
of beautiful brunettes, to be sure, but in my heart-of-hearts, I
considered myself a true gentleman who preferred blondes!

Things progressed for us, and by Christmas time the fires of
love were really burning. Sue returned home to her parents for
the holidays, and they weren't overly impressed that she was
getting involved with a "boy in a wheelchair." I even sent a
beautiful Bible home with her, to give them as a Christmas present. They thought it was nice, but seemed to downplay our

relationship.

When the school break came to a close, I met Sue at the airport in Salt Lake City. I was wearing a tuxedo-decked out to the hilt-and we went out to a romantic candlelight dinner. It was great, and the sparks were flying in both directions!

We dated steadily through the winter semester, and had the time of our lives. She was not only physically beautiful, but she had a little girl giggle and personality that was contagious. She also had the spiritual qualities I valued, and that was the clincher. In short, I was smitten from head to toe, and from all appearances, so was she.

> *"Many things in life will catch your eye, but only a few will catch your heart. Pursue those!"*
> ANONYMOUS

During the week of final exams, Sue began to consider where she would be living in the fall. She knew she had to put down a security deposit to hold a room, and she talked with me about it. This not-so-subtle maneuver caught my attention, believe me! I suggested that she consider some other place like Wymount Terrace, a married student housing complex near campus. This was quite a romantic way to drop a hint, but she got the message, and soon we were planning how she could introduce me to her increasingly skeptical family. We decided I should fly to her home to get acquainted, and get the mystery of the wheelchair boy out in the open. So she called her mother and told her how serious we were, and that she was bringing me home to "meet the family."

Sue didn't tell me this at the time, but I later learned that after getting the news that I was coming to meet the family, her mother cried for three days straight. She just couldn't accept the fact that her daughter was marrying a paraplegic. She was understandably afraid of the unknown-afraid for what her daughter might be getting into-afraid that she would have to do everything, yet never be able to bear children. I couldn't blame her, either. Even so, I really didn't see myself being different than anyone else-just doing things differently, that's all. I also had every confidence that I would one day become a father. I didn't know when, or how, but inwardly I sensed that the children we needed would be sent to us.

But things in Cleveland, Ohio, were not easy. Once out of my comfort zone, I had a difficult time doing things. While staying in her mother's home, I had to crawl upstairs, drag myself into bathrooms, and so forth. One time, in fact, Sue came to the rescue. Inviting me to wrap my arms around her neck, she hoisted me on her back and carried me up the stairs. I was worried about her hurting herself, but I was also enjoying the ride. Being that close to her was not a chore, and although her mother lifted her eyebrows when we did it, I was in seventh heaven.

Understandably, Sue's mother and father found our relationship hard to cope with. I didn't worry about it, though, because when Sunday afternoon rolled around, I did too-directly into the living room of Sue's mother's home. It was there that I proposed to her. I wanted to do wheelies when she accepted, but I contained myself, and almost before we knew it, we were engaged! It was totally awesome, and I had every confidence that her parents would come around the more they got to know me.

We've often laughed that we met on Halloween, became engaged on Easter, and got married on our anniversary. But then again, everyone gets married on their anniversary. We were falling swiftly in love, and although I was having a difficult time convincing my future in-laws that I could adequately care for their daughter, it was a time of great fun for both of us. Being with Sue was more and more intoxicating, and there was nothing I wouldn't have done for her. She made me feel like a million bucks!

Finally, on August 25, 1984, we were married in the Provo LDS temple. Sue's family traveled all the way from Ohio to be with us, and it was especially fun hosting her two younger brothers, Dave and Pat. That evening we held a reception in the Provo Excelsior Hotel, and over 1,000 guests attended. It was an exhausting four-hour celebration, especially for our parents, but they went with the flow; and before we knew it, we were on our honeymoon to San Diego.

And, I'll have to say, there never was a happier honeymooning couple in California! We went to the San Diego Zoo and Disneyland, enjoying the balmy, tropical climate and beauty of southern California. Mostly, though-as with all honeymooners-we took time for us, exploring emotions and perspectives, and growing even deeper in love. It was a time we will always cherish, and as for me, I could hardly believe my good fortune. I had captured the woman of my dreams, the most beautiful lady in the world. What's more, I knew she loved me unconditionally, and we were beginning a lifetime of laughing and loving. It was awesome!

Reflecting on how my relationship with Sue's parents and brothers has grown, I feel close to them all, and appreciate how they've grown to accept me. Even more, they respect and love me, and this acceptance means the world to me. As they know, there is nothing I wouldn't do for any of them, as well as for their daughter. Next to my relationship with God, marrying Sue has truly been the greatest blessing of my life, and our children are a monument to our love.

> *"The soul of a woman lives in love"*
> LYDIA SIGOURNEY

Looking back on our courtship and marriage, I have to really give Sue credit. Without knowing all of the issues of living with a physically disabled person, she accepted our love with a blind faith, and never once looked back. I marveled at her response to our love. Two people coming together have enough issues and scripts to deal with without fashioning a marriage around a wheelchair. Add this lifelong physical limitation to the mix, and a person really has to stop and look at things from all angles. But Sue accepted me lock, stock, and wheelchair. She knew she had what it took to make our marriage work, and so did I. In fact, I had no doubt but that we would build a family and home life that would fulfill our every need.

Sue adds: "Marrying anyone is taking a giant risk because we don't really know someone well enough to make such a commitment. Marrying someone who is permanently disabled involves taking a risk that is filled with uncertainty. And this is how it was with us. I wanted children more than anything in the world, and I wanted a husband who could direct the course of our family with consistent, gentle leadership. Although I thought this was what I had in Mike, I didn't really know the extent of his greatness. Nor did I understand the power of faith, and reliance upon the Lord, in fashioning a 'forever' family."

One of the issues we had to deal with was the expression of intimacy in our marriage. While this part of our relationship is very sacred and personal, we want to express where we're coming from. We have found, like others before us, that true intimacy is emotional, and of course spiritual in nature. The physical expression, by itself, is but a vehicle-a facilitator, if you will, to help us become "as one." As a couple, we have been fortunate in learning to be intimate in each of these ways, and our scripts and expectations are sculpted by our love and mutual admiration. How blessed I am to have such a giving companion, and to be encircled with her love. I just hope I can always give to her, and meet her needs, as she does mine. After all, from our perspective this is what marriage is all about.

CREATING A FAMILY SCRIPT

As we married, we learned to rely upon spiritual guidance, too . . .especially in our desires to procreate, then to rear a family. We learned to define our relationship in a way that was fulfilling and rewarding to us both, and with the help of medical science, we have been blessed in bearing and rearing three beautiful children. These include Matthew, age eleven; Megan, age seven; and McKenzie, age one. We love them fiercely. What's more, we even think they love us! There is no greater joy than giving birth to children, then directing them along life's highways and byways. We're having the time of our lives, and we can't imagine a life without these beautiful kids.

I remember the night Sue told me she was expecting our first baby. She was excited to share the news with me, and all I could say was, "Wow! That's scary!" I was totally blown away with the thoughts of becoming a father, just as Sue was in anticipating her role as a mother. Matthew was finally born, and while we felt overwhelmed with the unknown challenges ahead, we couldn't have been more thrilled.

> *"Public presentations are like babies; fun to conceive, but tough to deliver!"*
> ANONYMOUS

Even though I gradually matured into the role of being a dad, life has not been without its difficult "fathering" moments. Still, our kids have never thought of me as a "disabled dad." Of course there comes an age when they've realized that most dads are not in a wheelchair, but that hasn't fazed them. I do all the things fathers and their kids do together-including fishing and playing basketball, as well as playing family games such as Monopoly, Yahtzee and Clue-and although sometimes I have to do some things differently, they haven't seemed to mind.

Having two daughters is rewarding, as Megan and McKenzie both sit on my lap and comb my hair-among other father/daughter activities-and this gives me the greatest joy a father could desire. When Megan was six years old, I began having the impression that our family wasn't complete; that we were supposed to have another baby. These feelings persisted until finally I shared them with Sue. She was as surprised by them as I was. Subsequently, we returned to the doctor and announced our decision to try for one more child. We needed this fertility specialist's help, and we only had the funds for a few procedures in which to have Sue become pregnant. It had taken many more times than that for our first two children to come along, but we had faith to match our limited funds, so we pressed forward.

Two weeks passed, and sure enough, we found out that Sue was, indeed, expecting our third child. We were ecstatic! Nine months later, when McKenzie was born, we knew she was a gift from heaven. She is our third miracle baby, and like the others, I can't seem to have her near me enough. Speaking of my role as a father, as I continue to deal with my physical condition, Sue states: "One of Mike's greatest gifts is his ability to pick up on the pain of our children. It's almost like he has an antenna, and can detect when one of them is experiencing negative emotions. As I've analyzed this, I think this ability has something to do with how Mike feels, himself. He is in constant pain, and so is able to empathize at a higher level than most. The kids love him for it, because he is always able to help them deal with the hurt when they have an illness or injury."

While our children have been basically healthy, they have had tubes in their ears and common problems like that. In addition, Megan had to be admitted to the Primary Children's Hospital for ten days with a sinus infection. Other than these normal occurrences, we're a healthy, action-packed family. Our biggest problem is my speaking schedule, and having to be gone so often. This gives Sue the feeling of being a "single mom," but she doesn't complain. She is my biggest fan, just as I am hers, and the way we support each other really works.

In speaking of our children, I often tell Sue that each of them has qualities I would like. I'm not just saying this, but I am continually surprised with the miniature adult that surfaces in each of them. We swim together, go snowmobiling, camping and fishing, take rides, and do about everything a family could. We love establishing family traditions, and one of these is our annual Christmas Eve excursion to see the lights. We vote on the homes best decorated, and come up with a winner every time.

But, I have to say that one of the most enjoyable things I do with the kids is wrestle with them. They like it, too, since they can gang up on me and pin me quite easily. They're pretty good sports about it, though, and soon get off my legs so we can have ice cream, or whatever. We're not a perfect family by any means-we have our weaknesses, and our moments of frustration. Even so, we're able to work through them and somehow grow even closer in the process. Sue and I firmly believe in positive parenting, and we know our attitude will largely impact the types of adults our children will become.

6

STRIKING GOLD—
AN IMPOSSIBLE DREAM

> *"It is better to shoot for the stars and land in the trees,*
> *than to shoot for the trees and land in the mud."*
> ANONYMOUS

REACHING FOR THE STARS

Arriving at the ripe old age of twenty-five, I found myself loving to play basketball every bit as much as I did as a young boy. What's more, I had become pretty good at it. I loved to dribble, pass, and shoot, as I continued to sharpen my skills in these areas. Dribbling behind a wheelchair was a little more difficult than dribbling behind my back, but I soon was able to do it. It was likewise challenging to learn to shoot from a seated position, without being able to push off with my legs, but that too pretty much became second nature to me.

I've heard it said that you will miss one hundred percent of the shots you don't take, and I believe it. So I became a "shooting point guard," relying on my upper-body strength to get the ball through the hoop. I knew the game was identical to the college and pro game-ten-feet high baskets, a three-point line, fouling, and so forth. The only difference I could see was that we used rubber tires instead of Nike gym shoes. While not wanting to appear boastful, the level of play is also the same.

The skills of passing, blocking, and shooting are refined to an art at the highest level possible.

So it was with this background that I began to inquire about "trying out" for the U.S. Paralympic basketball team. The Paralympics was the second largest sporting event in the world, a sister organization to the Olympics consisting of athletes with physical disabilities. The Paralympics were held in conjunction with the Olympics, only one week later, and always in the same city. Anyway, I was pursuing my Masters of Business Administration at Arizona State at this time, and the tryouts were being held in nearby Tucson. The top sixty wheelchair basketball players in the United States were invited to try out.

I gave it my best shot (pardon the pun), and although I didn't really appreciate the odds against me at the time, I was good enough to make the team. I was the youngest player to be selected, and I was ecstatic in being able to compete in the Seoul, South Korea Paralympics.

As an interesting sidelight, when I went to try out for the Paralympic team, I found that not just paraplegic athletes competed. Some were amputees, while others had spina bifida and other physical limitations. Many of these great athletes don't live in a wheelchair, but use one only to compete. This was quite a paradigm shift for me that I'll talk about later.

> *"Only in the dictionary does success come before work."*
> ANONYMOUS

ARRIVING IN SEOUL

When we arrived in Korea, a bus was at the airport to escort us to the Olympic Village, and we were accompanied by armed guards. I had quite a lump in my throat, and I could hardly wait for our team to begin its first practice. The games finally began, and before I knew it we were in the gold medal game, playing against the highly touted Netherlands. Although I was the only rookie on the team, I was the starting point guard, and averaged about ten points a game.

During the half-time chalk talk, it really hit home to what a quality event I was participating in. This wasn't a second rate display of talent, but a game involving the finest athletes in the world. What's more, they were athletes who, in every case, had overcome tremendous personal difficulties before joining the team. I knew at that moment that I had finally realized my dream. I was playing for the greatest trophy offered, and what's more, I was confident we could win.

And win we did, with a final score of 55-43. I was happy to have scored eight points, but I was even more pleased with how well our team had performed together. Just as in life, with a corporation, or even a family, we had mastered the "team" concept. Selfishness and grandstanding had been eliminated, we had each played our role, and our team had achieved the highest level of success.

Twenty minutes after winning the gold, we rolled out to listen to the national anthem while we were each presented our medal. At that moment, I experienced emotions that are impossible to put into words. Only Olympic and Paralympic athletes receiving the gold medal can possibly comprehend it.

> *"Winning is overcoming that person inside of us who wants to quit."*
> ANONYMOUS

Reflecting on this singular experience of my first Paralympic competition, one of the most meaningful insights I gained was in watching other disabled athletes perform on this level. My sensitivity to them, as a whole, actually began in San Francisco, when the United States team was boarding the plane for Korea. Here I was, surrounded with 600 disabled individuals who had become athletes at the highest level possible.

Our experience at the airport was nothing short of a comedy. On other similar trips, I had seen it all-beginning with parking the van, followed by going through a metal detector and being frisked. My gold medal again triggered the alarm, and when I told the security officer about my medals, his reply was predictable. "Yeah, right," he scoffed. "And my name is Marilyn Monroe!"

Speaking of travel in general, I always try to show up late so the flight attendant, in an effort to avoid delays, will just have me sit in an empty "first class" seat. When this happens, I have a great flight, enjoying the bennies to my heart's content.

Then there is the point of arrival. When the plane pulls into its assigned gate, the flight attendant will welcome everyone to the city, then remind those who need "special assistance" to remain seated until everyone else has exited the plane. I always chuckle when this request is made of someone who spends his entire life "remaining seated!"

Returning to the Paralympics and my present flight, the next fourteen hours with these exceptional peers was an eye-opener I'll never forget. They were athletes with every disability imaginable. These included dwarfs, blind athletes, amputees, and persons who in one way or another had overcome a severe disability. In fact, seven of the twelve members of my basketball team carried their wheelchairs into the gym, jumped in, and began to play. Seeing these teammates, as well as the other 600 disabled athletes, was an expanded perspective accompanied with deep emotions, and I came away from the games with an eternally altered mind-set of my own condition. I was blessed beyond measure.

One of the first friends I made after arriving at the Olympic Village was a dwarf fellow from France. He had a great sense of humor, and because I felt taller than him, we hung out a lot together. He was an eighty-pound weightlifter who bench-pressed three hundred pounds. Another of my friends-who is also a dwarf-once told me that he originally had aspirations of becoming a professional football player. Then, realizing his size limitations, he adjusted his goals to simply becoming an NFL official-size football. He was great, and loved to have fun.

One day my Paralympic teammates and I decided to have a good laugh. So we encouraged our small French dwarf friend to climb inside a pillowcase in our dorm. We then hid behind the corner while the maid came to clean the room. When she got to the pillows, she "shook him out," and was she ever surprised to see him! She didn't speak English, but did she ever express herself!

We all laughed, our dwarf friend jumped down off the bed to join us, and we were off to have a great day. Those moments will last forever in my mind, and were as much a part of the Paralympics as was the competition.

By speaking of the athletes' various disabilities, this is not to infer that we were participating in the Special Olympics, the games for the intellectually disabled. Their games are held separate and apart from ours, and I enjoy watching them compete.

Let me briefly share two of the most poignant moments in the 1988 Seoul Paralympics. If you can believe it, I watched a man with one leg high jump seven feet. If that wasn't spectacular enough, I then watched a team of blind athletes play a game called goal ball. Similar to soccer, the ball was electronically programmed to regularly "beep" so that the team members could locate the ball and kick it. This experience was absolutely incredible-and I thought I was athletic!

> *"Olympic success is always preceded by Olympic motivation."*
> ANONYMOUS

SHARING WITH OTHERS

After I returned home, my mom was so proud of my gold medal that she wanted to have it bronzed! Bronzed? I could hardly believe it; but I talked her out of this nonsense, and she later had it encased in a gold display frame. I didn't really care for this, though, so I took it out and carried it around with me. I didn't do it to show off, but simply to let young kids touch it, seeing what they could possibly do if they worked long and hard enough. I simply wanted to share this unusual moment in history when I had played with the best wheelchair athletes in the world, and this was my way of doing it.

Still, in spite of winning this medal, my greatest memories were of the journey to the Paralympics, rather than of taking the gold. It was one of the most intense and action-packed journeys imaginable, one that I'll never forget.

MY SECOND PARALYMPICS-WITH A TWIST OF FATE

Four years after competing in Seoul, I found myself again on an airplane. Only this time I was traveling with my teammates to Barcelona, Spain. It was to be a competition of destiny, only it wasn't the kind of destiny I had hoped for. The games in Barcelona had been heralded by the media as the best Olympics and Paralympics in the history of the games.

Things went our way once again, and finally we found our-selves accepting gold medals while the national anthem was playing and the flag was raised. We had once more gone all the way, again beating the Netherlands by three points. Little did any of us expect that not long after we had returned to our homes, we would receive a letter in the mail stating that we had to give our medals back, or we would be barred from competition for life.

The Paralympic officials had randomly selected one of our star players for a drug test, and he had tested positive. It was a painkilling drug that, while illegal to use at the time, is now legal for certain athletes to use in Paralympic competition. None of us suspected a violation at the time, so sending that medal back, after having worked our hearts out to receive it in the first place, was one of the most difficult things I have ever done.

Perhaps I should add that, as a team, we petitioned the committee, then went through the courts, in vain, all in an effort to have the committee be reasonable. Even so, the law was the law, and even though this player hadn't taken "illicit drugs," per se, we still had our medals stripped from us.

While I don't harbor negative feelings with this friend and teammate whose actions inadvertently caused us to lose our gold medals, still I feel badly that one man's actions in two minutes overruled several thousand hours of work and sacrifice to achieve a goal, and that hurt. So, while this was initially a negative thing in my life, I have since used it as a tool to learn from, to dig even deeper in resolve so that I could again capture the dream I originally had. I've also come to appreciate that if life is always "rosy," and if we never lose anything, we will likewise never appreciate the good things that happen.

I have purposely told this story without using this friend's name. I didn't want his reputation tarnished in any way. But now I would like to tell you who he is. His name is Dave Kiley, and he is considered by all to be one of the greatest wheelchair basketball players ever. Nor does the story end here. Following the United States team being stripped of our gold medals, Dave spent hundreds of hours of his time trying to have the committee's decision reversed. He was devastated when this didn't happen, as were we. But still the story does not end here. Dave went on to become the Commissioner of the National Wheelchair Basketball Association (NWBA). He is a great example of using positive energy, rather than negative, when something unexpected and undesirable takes place.

As an indicator of how Dave has helped our sport to grow, in February of 1998, he lobbied the National Basketball Association, and for the first time ever we were invited to Madison Square Garden, in New York City, to compete as part of the NBA All-Star weekend. Our game was held on Thursday evening, and as a team we had the time of our lives. There were several thousand spectators, and the West team beat the East team by a score of 78-73.

> *"Not what men do worthily, but what they do successfully, is what history makes haste to record."*
> H.W. BEECHER

MY THIRD PARALYMPICS

Reflecting on my seemingly never-ending basketball saga, I made the Paralympic wheelchair basketball team again in 1996. This time the games were to be held in Atlanta, Georgia. Because of my previous two competitions, I was going as a veteran, and I was going to have fun! While obtaining a gold medal was my personal goal, the Paralympic experience had grown way deeper than that. From experience, I knew that rubbing shoulders with, and learning from, the great athletes of the world, would give me perspective and a further insight to myself, as well as to my life's quest for excellence. What's more, although Sue had gone to the other two Paralympics with me, this time we would be taking Matthew, and that was something special. He wouldn't remain for the entire games, of course, but it was an experience we would remember as a family, and this gave an entirely new meaning to it.

What I didn't know until competition time was that I was about to learn a lesson of an entirely different color. We progressed undefeated, pretty much as we had expected to. Then, in the semi-finals, we played Australia. We knew that if we beat them, we would play Great Britain for the gold the following evening. While we led through much of the game, one of Australia's players, Troy Sachs, had the game of his career scoring forty-two points, and we ending up losing by a single point. It was nothing less than horrible, for us and most of the 10,000 spectators, especially since we knew we should have beaten them!

Following the game, the locker room was pandemonium! Wheelchairs, both empty and with others sitting in them, were flying everywhere-along with cursing and yelling and unbridled anger! It was a scene I'll never forget! We just didn't know how to lose; but because I was the only player on the team who had also competed in the previous two Paralympics, I could understand the response. Our coach, Brad Hedrick, finally gave a very emotional speech, and that seemed to settle our emotions and prepare us for our final game.

This meant that we would play Spain for the bronze medal, and we won this game quite handily. It was different taking third, but it was also very gratifying. The one consolation in winning a bronze medal instead of a silver medal is that at least the players return home having won the final game.

Since receiving this medal, I've had fun showing it to people, especially the braille inscription on the back for the blind athletes. While rubbing this inscription, I tell people that it says something like "Australia kicked your butt. Better luck next time!"

> *"Be a winner, not a whiner!"*
> ANONYMOUS

During the medal ceremony, I honestly didn't have animosity toward the team that had beaten us the day before and was now accepting the gold. They deserved to win, and had done it fair and square. In a way, I was happy for them because I had sat in their chairs twice before, and I knew what they were experiencing. I also knew we were winners, even with a bronze medal. Winning any Olympic or Paralympic medal was the dream of a lifetime, so knowing my family was there to celebrate this moment with me made it all worthwhile. Although the bronze is certainly nothing to be ashamed of, I will always believe that if we had paid just a bit more attention to detail, we would have and should have won the gold. I guess one never knows, and that's why the game is played. Even so, deep inside-and if all went well-I knew I'd be there to give it one more try four years later!

AN UNPRECEDENTED FOURTH PARALYMPICS
As 1997 came to a close, I found myself traveling to the Olympic Training Center in Colorado Springs, Colorado, where I again tried out for the team to play in the world championships. I lucked out again, making the team, and I was no less thrilled then than I was the first time. I would now be competing in the world championships, in Sydney, Australia. After these games, I would hopefully be training for Paralympic

competition in the year 2000, to be held at the same location. What's more, because the Aussies had beaten us in Atlanta, I have an even greater motivation to go and beat them on their home court.

If I am able to make the team and participate, I will be one of only two or three wheelchair basketball players to have competed in four Paralympics. This is not to infer that I am trying to be better than others, for that couldn't be further from the truth. But I have had the inner goal of participating in four Paralympics for as long as I can remember. I simply want to achieve my greatest potential as an athlete, just as I've determined to do in life-both as a professional and as the head of the greatest family ever!

> *"There is no nobility in being superior to another person. True nobility is being superior to one's past self."*
> BRENTON G. YORGASON

On a personal note, I find it intriguing that I wear No. 12 on my jersey. Why? Because this is the same number worn by John Stockton of the Utah Jazz. Since I'm from Utah, and because I'm basically the smallest player on the team-and since our facial features are similar and my hair is dark and cut like his-many people have compared me to him. Interestingly, I play for a team in Utah called the Wheelin' Jazz. My friends and I started this team eight years ago, and we have been consistently ranked in the top twenty nationally.

As point guards, John Stockton and I have similar assignments. It's our job to create the plays, pass off, and harness the individual talent that comes together on the court. Of course it's giving me way too much credit to compare myself to John Stockton, but I have thought of letting John adopt me just so I can be part of his inheritance! Just kidding. What I would like to do, if given the chance, is teach John how to dribble behind his chair. Then he could do it all-dribbling behind his kitchen chair as well as behind his back on the court!

LOOKING BACK

Reflecting on my own experience, and how my game of basketball had evolved, I actually began playing competitive sports in the community Bantam League. I played baseball, basketball, and football, and loved them all. What made things even more fun, as I got into basketball, was that my dad was always my coach. He was great, too, a real people builder, and with his encouragement, I knew I could do anything! It was awesome! What's more, my mom was always at my games, cheering me on. Both parents had their own way of motivating me, and I would have died trying to perform to their expectations.

I need to mention that I wasn't just a pampered kid who spent his summers attending one basketball camp after another. Instead, my dad ingrained within my mind the need to practice hard all the time so that I could get ahead of my competition. His work ethic became mine, and I knew only one speed-fast forward!

> *"That which we persist in doing becomes easier to do; not that the nature of the thing has changed, but that our ability to do it has increased."*
> EDWIN MARKHAM

To illustrate how intense I was, I remember many times getting up in the wintertime, shoveling snow off our neighbor's driveway, then putting on my gloves and shooting baskets for an hour to an hour-and-a-half. This I did in the early morning, prior to heading off to school. I really did want to get ahead of my competition, although I like to think that I also wanted to be the best "Mike" I could possibly be.

Part of my motivation to excel came from a kid in the neighborhood by the name of Kirby Johnson. He was not a particularly gifted athlete, but he would literally practice basketball four or five hours a day. He was consumed with the desire to play in the NBA, and his work ethics showed it. The best part of this for me was the fact that I always had someone to play basketball with. Because my dad was a former high school coach, I had the basics down, so spending so many hours with Kirby turned out to be a great benefit to me.

To summarize my life's experience with competition, I believe I have unquestionably accomplished more because of it. To be even more explicit, I have always tried to compete with myself, rather than with others. Competing, in and of itself, is rather shallow. . .if all a person wants to do is "beat out" their competitor. Challenging oneself, then having the dogged dedication to train at one's highest possible level, is really what makes it worthwhile. After all, this is what makes me keep going, and make the sacrifices necessary for one more Paralympics. I have the faith that I can once again train and compete on that level, and so-God willing-I'll give it my best shot!

7

SECURING SUCCESS THROUGH ADVERSITY— SIX BULLETPROOF PRINCIPLES

> *"If we resist change, we'll fail,*
> *if we accept change, we'll survive;*
> *and if we create change, we'll succeed."*
> ANONYMOUS

Throughout my life-and especially after my accident-I've had a penchant for not exposing my weaknesses. I'm sure being in a wheelchair has made me more sensitive to this than I may have been otherwise, but nonetheless it's true. I have a yearning that comes from deep within, and it is to succeed at the highest levels. While I am limited to what I can accomplish, physically, I have never felt limited as to what I can do overall. If anything, people have called me an over achiever. I've accomplished whatever I've set out to do, and that includes serving a full-time mission for my Church, graduating from college, receiving an M.B.A. at Arizona State University, and being a successful inspirational speaker and businessman. For the past five years, I've worked full-time for Intermountain Health Care (IHC), in Salt Lake City, Utah. At this time, we have created a new division at IHC that incorporates the principles of Attitude Therapy.

I've not meant to share these accomplishments in a boastful

manner, for if anything, I give credit to Sue and others-and of course, to God-for supporting me along the way. I fervently believe in the power of the mind, and in living with great expectations. To accomplish this, I have developed six bullet-proof principles I would now like to share. After all, these have become my "keys to freedom," and to enjoying a successful, action-packed life.

SIX BULLETPROOF PRINCIPLES FOR SUCCESS

Principle Number One: Live to grow and to change, rather than merely to "exist." In other words, tackle life with a passion! I first realized this need almost immediately after I had been shot. At that time, my lens was cloudy, and I had to find a suit-able "mental Windex" to regain clarity of vision. This clarity literally began with deciding to get up every morning. There was no pressure to do this, so I had to dig deep down inside and create the pressure. Becoming passionate was the result. It was mind over mattress of the first order!

The opposite of living with passion is simply giving up-allow-ing someone to push our "kill" button. Many people die at age seventy-five who stop living at age thirty. From my experience, this is a disease fostered by low self-esteem.

When I was younger and newly injured, I began competing in marathons. I loved the challenge, and although I always fell behind at the beginning uphill portion of the race, I passed most of the runners on the downhill stretch. One day, while competing in the Deseret News Marathon, I realized that par-ticipating in a lengthy race like this was much like life. At the beginning, people are there to cheer us on and encourage our efforts. Likewise, thousands are there to cheer us across the fin-ish line. The difficult part-the twenty-four miles of asphalt in between-is where the measure of a man is made. Although we always appreciate support from others, in the final sum of things, our own decisions determine our outcome.

Principle Number Two: Learn to take responsibility for our attitudes and our actions! That is, rather than project blame, or to allow the seeds of anger and discontent to enter our hearts when something goes wrong, we should be "in control" of our emotional response. The first cousin to this idea is the need we have to set goals-to create a life map. After all, corporations have mission statements, so why shouldn't individuals. The person who expects nothing from him or herself will certainly never be disappointed.

A recent example of this notion is the disabled pro golfer, Casey Martin. Because of Casey's extremely atrophied leg, a judge in Eugene, Oregon recently made it possible for him to ride a cart on the pro golf tour. As reported in The Daily Herald[3], "For Martin, the landmark ruling was not just a victory for him; it was a symbolic victory for all those with a disability who've been told they can't. 'I realized if I win, [Martin was quoted as saying] it would open the way. That's something to feel good about.'"

Not only has twenty-five-year-old Casey Martin excelled in his sport by winning a PGA tournament just weeks before this court ruling, in Lakeland, Florida, but he is living within his limitations while invoking the Americans with Disabilities Act to allow him to compete on the professional level he has worked so hard to achieve. It is a great victory for one who has taken responsibility for his own future.

Principle Number Three: Live not only to accept adversity, but to relish the acceptance of it. After all, the finest steel is made in the hottest furnaces. I call this strategy welcoming the "refiner's fire."

---

3. February 12, 1998, p. B1

A year following my accident, the friend who shot me was hanging out at the local mall with me, having a good time. Even so, I could tell that everyone who passed looked at me with a curious eye. At the time, I didn't like it, and I wanted to put a sign on my chest that read: *I got shot, you geek! What's your problem?*

Although it may seem contrived, we actually had a lady come up to us and ask me why I was in a wheelchair. My response was awkward, to be sure, but I simply and truthfully said, "Because my friend here shot me." He and I laughed, but that lady surely didn't.

After returning home that afternoon, I began to realize that people weren't being rude in looking at me or asking about my condition. They were simply being curious. I then realized that the way I carried myself was going to be the way they would regard me. They would either see me as a person with a personality, or as a paraplegic who was impaired. It was at this moment of discovery that I determined to not only accept my adversity, but to relish the acceptance of it!

A later experience I had that let me know how I had accepted my mishap was in speaking with the Utah Jazz announcer and former NBA star, Hot Rod Hundley. In talking about his golf game, and his personal "handicap," he asked what my handicap was. Grinning widely, I simply said, "PAR-a- plegic!" We both laughed, and again I was reminded of the importance of a healthy, humor-filled accepting attitude.

Principle Number Four: Live with a "Service Mentality." From what I've seen, life is made up of takers and givers-those who scavenge from society, and those who contribute. While providing a living is important, what's even more essential is to serve others. Today's hedonistic society is all too selfish, and that's frightening. If I can be remembered for anything, it would be that "Mike cared."

As an adjunct to this idea is serving others "positively." Ken Blanchard, a popular training and development instructor once wrote a one-minute course in public speaking. In it, he states, "Of all the concepts I have taught over the years, the most important is about 'catching people doing things right.' There is little doubt in my mind that the key to [influencing] people is to catch them doing something right, then praising them for their performance."

I've heard it said that no one cares how much we know until they know how much we care. We care by lifting and genuinely praising others, rather than confronting and condemning them. This, after all, is the only way to serve!

Principle Number Five: Live *for* a family, rather than just "with" one. The greatest success I've had, of course, is with my family. Both Sue and I feel that the family-rather than the individual, or the village-is the basic unit of society, and that greater joy can come from this experience than from any other episode life could present.

Permit me to share a memo I recently received-reporting a true story from the Vietnam war. I call it *"packing parachutes."*

Charles Plumb, a U.S. Naval Academy graduate, was a jet fighter pilot in this war. After seventy-five combat missions, his plane was destroyed by a surface-to-air missile. Plumb ejected and parachuted into enemy hands. He was captured and spent six years in a Communist prison. But he survived, and is alive to tell about it today.

One day, when Plumb and his wife were sitting in a restaurant, a man at another table came up and said, "You're Plumb! You flew jet fighters in Viet Nam from the aircraft carrier Kitty Hawk. You were shot down!"

"How in the world did you know that?" asked Plumb

"I packed your parachute," the man replied proudly. Plumb gasped in surprise and gratitude. The man pumped his hand and said, "I guess it worked!"

Assuring him, Plumb replied, "It sure did. If your chute hadn't worked, I wouldn't be here today."

Plumb couldn't sleep that night, thinking about that man. Of this night, he states, "I kept wondering what he might have looked like in a Navy uniform-a Dixie cup hat, a bib in the back, and bell bottom trousers. I wondered how many times I might have passed him on the Kitty Hawk. I wondered how many times I might have seen him and not even said, 'Good morning, how are you?' or anything because, you see, I was a fighter pilot and he was just a sailor."

Plumb thought of the many hours the sailor had spent on a long wooden table in the bowels of the ship, carefully weaving the shrouds and folding the silks of each chute, holding in his hands each time the fate of someone he didn't know."

When I think of Plumb, I think of how spouses and parents spend their lives packing parachutes. We spend our energies providing the means for members of the family to make it through the day. It is not only our stewardship, it is our opportunity and blessing-in fact, it is our greatest opportunity and blessing!

Principle Number Six: Learn to laugh! Life is so serious, with such great demands, and we all need to lean back, stretch, and laugh!

One humorous incident took place when I was in college, and before I met Sue. My friend, Doug, and I wanted to spiffy up a blind double date, so I loaned him one of my wheelchairs and we both went "wheeling" up to Deseret Towers to pick our dates up. We were not only blind dates to them, but were also disabled!

The girls couldn't believe we were both disabled, but they went with us to the university bowling alley. On our way down the hill, tragedy struck! Being unpracticed in the art of wheelchair riding, Doug lost control of his chair and landed directly in a window well. People came to his rescue and pulled him out, then we continued on to campus. Once there, this fellow continued his charade by bowling from his chair, just as I did. This was okay until he again fell out of the chair.

Before the girls could respond and run to his rescue, he jumped to his feet, yelling, "Surprise!" They could see they'd been duped, but soon all was forgotten and we had a great time. We're not sure our two dates liked it, but we certainly did.

Another humorous situation that illustrates how one can actually have fun, even with what he has lost, is reflected in the antics of one of my Paralympic basketball teammates, Reggie Colton. He is a double amputee, and sometimes when we pre-board a plane, he'll hoist himself into one of the upper luggage bins. He is one of the strongest men I know, and just as agile. We then close the lid, only to have it opened by a startled passenger moments later. As they open the compartment, Reggie jumps out and frightens them. Maybe this is taking things a bit too far, I don't know; but I do know that in the face of personal loss a person can have a much easier time of it if they are able to laugh at themselves and their situation rather than to feel sorry for themselves.

PERSONALLY SPEAKING

Perhaps the greatest secret to the vitality of our marriage has been our ability to share feelings. Not simply to express thoughts, as most couples do, but to get down to the nitty-gritty, and express how we feel about something. I'm sure my disability has contributed to our need to communicate-another one of those unexpected benefits-but nonetheless, we've grown close because we talk things out, and that's made all the difference. We share, and we care. Oh, marriage hasn't been without its moments, that's for sure. In fact, for both of us it's the hardest thing we've ever tried to do. But we work as a team perhaps because we've been forced to do so. This is a nugget of gold that we've discovered, and we wouldn't trade it for anything.

We have found it intriguing, over the years, that people want to focus on the more personal part of our marriage. Whenever we meet someone, we've almost come to expect that they have two questions: First, can we have sex? And second, are our children ours? We're not defensive about these questions because we know the inquisitive nature of human beings. Sensing their thoughts, and listening at times to their queries, we politely explain that we couldn't be happier in our marriage. We then say, "Yes, they're our kids. I just look at them and rejoice that they look more like Sue than they do me!"

But our success, both as contributing adults to society, as well as a husband and wife team that is doing its best to rear children with values and aspirations, is measured by the joy we experience each day that we live. We're home bodies, and we love it, even though we're always on the run, looking for things to do and friends to do them with. Although my speaking engagements and sports competitions force me to travel a great deal, we've adapted to this regimen, and live for the times we're together. This, after all, is the greatest success a person can experience.

EMPOWERING LIFE'S PASSIONS—
THE PERKS OF THE CHAIR

> *"In the history of passions, each human heart is a*
> *world in itself; its experience can profit no others-"*
> ANONYMOUS

THE UNEXPECTED PERKS OF "THE CHAIR"

Other than the self-serving perks such as being privy to handicapped parking stalls and avoiding lines at Disneyland, there are other more relevant benefits to "doing life" from the padded seat of a wheelchair.

When I was first placed in a chair, it was two weeks after my accident. After several days of getting stabilized, I was strapped onto a tilt table to help me regain my equilibrium. The physical therapists would first lift me into a forty-five degree angle, then put me back down flat. They'd then repeat this process, taking me to an even steeper angle. Gradually my brain adjusted, and I was able to sit up without getting dizzy.

Following this procedure, I was placed in my first wheelchair. Although I simply sat in it, allowing my aid to push me down the hall to physical therapy, still it felt good to again be mobile. Not long after this, I was able to roll on out of the hospital for a few moments, enjoying the crisp fall Utah air.

At last, a few weeks after my accident, I was able to leave the rehab center for a day. My "coming out" party, so to speak, was going to my Uncle Kent's home for Thanksgiving dinner. I loved the dinner and the ball game on television, but my focus was on practicing wheelies and other maneuvers over in the corner of the family room. My competitive nature was again taking over, and I wanted to be the best wheelchair navigator in the world. Little did I realize at that early hour that spinning around in my wheelchair would become one of my favorite things to do. Definitely a perk of the first order!

> *"If you always do what you've always done, you will always get what you've always got!"*
> ANONYMOUS

Another of the initial perks of being in a wheelchair was being able to use my newly limited physique to get out of doing things. Music was an emphasis for our family, and prior to my accident, I had spent several years taking piano and guitar lessons, and doing okay with both. However, good fortune- or bad, depending on how you look at it-prevailed since I quit taking piano lessons at the time of the accident. I convinced Mom that I couldn't use the pedals, so that ended that. Seriously, I appreciate my folks giving me this musical background, and although I didn't become a professional singer like my little sister, Collette, I now see where my folks were coming from, and I want the same for my own kids.

A further perk that I've mentioned in a different context is the sympathy I got from my friends. One of my first discoveries was how interested girls seemed to be in me. I probably took things more personal than I should have, but I thought I was pretty cool doing wheelies, twirling around, racing up and down the halls, and in general just showing off. I now realize

how immature that was, but at the time, it gave me great inter-
action opportunities with my friends, and I took advantage of
every one of these opportunities.

THINKING DIFFERENTLY

One of the real traps people get into is the trap of sameness.
Perhaps the first thing my accident did for me was force me to
think differently. I've often heard it said that we all spend our
lives adjusting. In my case, I began to think that, contrary to
popular belief, my wheelchair was not my identity. It was simply another tool, like my glasses and my automobile, that could
help me perform at a level higher than I might otherwise have
been able to do.

This "different thinking" has reached into other areas of my
life, as well. Since evolving and healing from my accident, I
found myself avoiding judgment of others according to outward symbols. I look inside a person's heart, as I have stated-at
his or her motives, desires and attitude.

What's sad is that so many disabled people spend their lives
reacting to their perception of what others think of them,
rather than purposefully "doing their own thing." They won't
admit to being disabled, nor will they hang around others with
similar limitations. When I meet these people, I study their
statements, gleaning all I can from what they have learned from
their mistakes or misfortunes.

One of the special people who does his own thing is Mike
Johnson. Mike is a double amputee who lost his legs and several of his fingers in a land mine explosion in the Vietnam war.
He lived several miles from me, and after I got used to my
wheelchair, he kept pressing me until he got me to come out of
my shell and play competitive basketball. In retrospect, he was
one of the most important people who helped me overcome
adversity with class and with dignity.

CHEER UP, THINGS COULD BE WORSE!

Well, I cheered up, and sure enough. . .things definitely got worse!!

A year or so following my accident, I found myself in an arm wrestling competition with Coach Fuller, my weight-lifting coach at Orem High School. We were in his social studies classroom, and my girlfriend at the time was watching us. We went all out, and before I knew it, he had won-breaking my arm and leaving me racked with pain. My funny bone was broken, and was pulled all the way up to my wrist. However, I found it anything but funny! I thus spent the next two months with a cast extending from my wrist to my shoulder.

Not only was this injury very painful, but it greatly limited my ability to get around. Although I hate to admit it, I went "in circles" for these two months, getting nowhere. Just kidding. Truly, I was slowed down, but this interruption in my progress caused me to learn entirely new dimensions about patience, as well as being recipient to the "law of the harvest." I showed off and became stubborn, and I paid big time for it.

> *"The best way to predict our future is to create it."*
> ANONYMOUS

DEFINING A WINNER MENTALITY

When I was ten years old, I played on the Dodgers Little League baseball team, and we won the city championship. This was the first time I thought of myself as a real "champion," and I'll never forget the feeling. I was the catcher, and my big brother Scott-who was without question the best baseball player in the city-was the star pitcher. We were a duo that couldn't be denied, and I think that the love and mutual respect we have for each other, as adults, was actually put in place way back then. Having a competitive older brother can bring pleasures and perils. Such was the case with Scott and me. A year older than me, I always found myself striving and stretching, extending myself beyond my limits, just to keep up with him. He was a great teacher, too, and took time to lift me to his level. Because of this, and because he included me in his activities, I will always have a special place in my heart for him. We had sibling rivalry, to be sure, but when things all boiled down, much of my progress-both as an athlete as well as a person-I owe to Scott.

The feeling of being "the best" accelerated even more when both of us made the All-Star baseball team. I wasn't the biggest or the best, by any means, but I was determined! More than any other thing, this determined mind-set was the thing that helped me adjust after the accident.

I knew that knowledge was power, and that I had to seize the moment and get and then use the best education possible. The notion of acquiring and using knowledge is supported in a recent subscription letter written by Peter R. Kann, publisher of The Wall Street Journal:

"On a beautiful late spring afternoon twenty-five years ago, two young men graduated from the same college. They were very much alike. Both had been better than average students, both were personable, and both. . .were filled with ambitious dreams for the future.

"Recently, these men returned to their college for their twenty-fifth reunion. They were still very much alike. Both were happily married. Both had three children. And both, it turned out, had gone to work for the same Midwestern manufacturing company after graduation, and were still there.

"But there was a difference. One of the men was manager of a small department of that company. The other was its president. What made the difference? It wasn't a native intelligence or talent or dedication. It wasn't that one person wanted success and the other didn't. The difference was in what each person knew and how he made use of that knowledge."

CHOOSING THE "RIGHT" FRIENDS

When I graduated from elementary school and moved on into junior high, I attended Lakeridge Junior High School. When I entered my very first classroom-Arts and Crafts-I met Roger, who was mentioned earlier. I sat down and looked around, and eyeballing him, I said, "Hi, my name's Mike." "I'm Roger," he replied, "Roger Dayton." We shook hands, cementing the beginning of a friendship that has continued to this day. He and his parents, Gary and Zona, were especially supportive of me after my accident, and I owe them a great deal.

There wasn't anything Roger wouldn't do for me, nor me for him. Not only that, but I always knew where he stood on things. I knew he had values and standards, and that being with him would motivate me to stand for the same things.

My other closest childhood friend was Doug Jensen, who went with me to college, and who was with me when I first met my wife, Sue. While Doug was a good athlete-especially as a tennis player-this sport didn't consume his life. Together we would hang out, play chess or table tennis, or just chase around with high school and college buddies.

Other friends were of Doug's same quality. What I appreciated more than anything was the fact that these associates still wanted to be my friends! This realization came slowly, and sank deep into my heart. Other than my family, these friends were my cheerleaders, and they loved me for who I was on the inside, rather than being stymied by my outward limitations. Ofttimes, at lunch, my friends and I would leave school. We'd throw my chair in the trunk of whoever's car we were driving, and we'd head to McDonalds. I remember one time these friends jumped out of the car and began ordering their Big Macs, without realizing that I was still in the car. This experience actually made me feel good, knowing that my friends thought of me as "normal," and were not too consumed with my chair.

WHERE FROM HERE?

In Intermountain Health Care's rehabilitation centers, we stress the notion that individuals with disabilities should hope for the best, yet prepare for the worst. I take this motto seriously, and have great hope in my own future progress. Medical scientists are already having success in regenerating nerves, and in having the spinal cord repair itself. I don't think it's far-fetched to envision a day in the not too distant future when my son and I will walk together as we tackle our favorite golf course or fishing hole-or that I can dance on my feet with my daughters at their weddings. This is my hope. . .if not for me, then for others who will follow.

LIFE'S OTHER ADVERSITIES

It's interesting that, when speaking to me, people are often consumed with my injury-as if my disability was the only difficulty I had to contend with. These people are often surprised when I tell them that, just like themselves, I deal with all types of difficulties. Life is difficult for everyone, not just someone who lives in a wheelchair. Depression can set in when a business deal goes sour, anger and frustration can arise when others don't meet our expectations, and so forth. I often paraphrase the bumper sticker which states: *Life's tough and then you die!*

Everyone has heartaches and disabilities-they are simply more obvious for some than others. In fact, from what I've seen, often the emotional and spiritual disabilities a person experiences far exceed any physical disability I might have. This is perhaps true because when others can't see one of these inner disabilities, a person has little incentive to overcome them. In fact, these weaknesses often become part of the baggage retained and even "made sacred" as they are kept in a theoretical fanny pack. People take them out and display them when convenient, then quietly tuck them back out of sight as they continue their trek-all with the extra weight this excess baggage provides.

Perhaps a perfect example of this would be my friend who initially shot me. While I immediately forgave him-and meant it-to my knowledge he never forgave himself. He grew up, married, and tried to do life by his own rules. Then one day he broke the law, and is now suffering greatly for his wrongs. In essence, he has at least temporarily lost all that he had worked for-including his wife and family. If he hasn't already, my prayer is that he forgive himself for the part he played in my accident. I hope that one day he'll realize how good a person he is, and that he can exist in society as a contributing, happy individual.

I firmly believe that, regardless of what people do to self-destruct, they are all looking for happiness. My nickname as a youth was "Happy Schlappi," and I wore the handle proudly. That is, until I broke up with a girlfriend. Then that particular girl would label me, "Crappy Schlappi." During those moments I was pretty down on myself.

> *"Trying to find happiness from the 'outside in' is a futile task. The only true happiness comes from the 'inside out.'"*
> ANONYMOUS

It is imperative that, along with all other habits, I internalize habits of self-progress, rather than those of self-destruction. In

her Autobiography in Five Short Chapters, Portia Nelson provides a humorous yet insightful sequence that sums up the core
challenge for attitude therapy. I share it as follows, as a setup
for success-or failure-depending on our ability to proactively
play life's game with the cards we are dealt:

Chapter I
I walk down the street.
There is a deep hole in the sidewalk.
I fall in
I am lost. . .I am helpless
It isn't my fault.
It takes me forever to find a way out.

Chapter II
I walk down the same street.
There is a deep hole in the sidewalk.
I pretend I don't see it.
I fall in again.
I can't believe I am in the same place,
but it isn't my fault.
It still takes a long time to get out.

Chapter III
I walk down the same street.
There is a deep hole in the sidewalk.
I see it is there.
I still fall in. . .it is a habit.
My eyes are open.
I know where I am.
It is my fault.

I get out immediately.

Chapter IV
I walk down the same street.
There is a deep hole in the sidewalk.
[Because of my inner attitude] I walk around it.

Chapter V
I walk down another street!

It is crucial that we sow seeds of constructive habits, and that we internalize the personal and professional integrity to be responsible for self. We thus have the strength of conviction to turn around, and to change directions to avoid getting bogged down in the quicksand of self-defeating attitudes and behaviors. Personal Change *is* possible, and can become not only a habit, but a core for one's character. This, in the final sum of things, will allow us to reap a destiny of success, wellness, and peace. After all, it is what makes life worth living! Hopelessness and failure are the greatest disabilities, and combine to introduce a plague that is stretching far and wide. Hope and success, on the other hand, breed upon themselves and allow a person to enjoy a life of accomplishments, self-fulfillment, and most of all peace.

> *"Change is constant, it's everywhere. The only change I don't like is changing my one-year-old daughter's diaper!"*
> ANONYMOUS

While I have enjoyed lofty athletic ambitions and achievements, my goal at this time in my life is to take my message of

Attitude Therapy and share it with the world. It is a remarkable high to be invited to speak to corporate America, and to share my story. I speak with tongue-in-cheek when I say how great it is to receive a standing ovation-after all, I can't stand and join in! Seriously, in my mind I not only stand, but I race forward with an enthusiasm and energy that could not be greater! After all else, what could be more meaningful and empowering? Enough said.

"A bend in the road is not the end of the road unless we fail to make the turn." — ANONYMOUS

"No matter what we have done up until now, our future is spotless." — ANONYMOUS

"How many parents, in moments of anger, push the "kill" switch of one of their kids by telling their little boy or girl that he or she will never amount to anything? How many kids then spend a lifetime working very hard to make their parent's prophecy come true?" — OG MANDINO

"Tell me about a man who has won the lottery, then about a man who has broken his neck, and you still haven't told me anything about either man's happiness." — ANONYMOUS

"Nothing is really work unless you would rather be doing something else." — SIR JAMES BARRIE

"If you have a big enough WHY in life, you can always discover the HOWS!" — ANONYMOUS

"Great things are accomplished by those who know how, but they will always be led by those who know why." — ANONYMOUS

"So often we seek for a change in the situation when all we have to do is change our attitude." — ANONYMOUS

"The same refinement which brings us new pleasures, exposes us to new pains." — RICHTER

"The best predictor of future behavior is past behavior." — ANONYMOUS

"All motivation is self-motivation." — ANONYMOUS

"Let challenges make you a better person, and not a bitter person." — ANONYMOUS

"If your lot in life feels empty, build a service station."
— ANONYMOUS

"You can make more friends in two weeks by becoming interested in other people than you can in two years by trying to get other people interested in you." — ANONYMOUS

"Do all the good you can
By all the means you can
In all the ways you can
At all the times you can
In all the places you can
To everyone you can."
— ANONYMOUS

"What an enormous magnifier is tradition!" — CARLYLE

"Many things in life will catch your eye, but only a few will catch your heart. Pursue those!" — ANONYMOUS

"The soul of a woman lives in love" — LYDIA SIGOURNEY

"Public presentations are like babies; fun to conceive, but tough to deliver!" — ANONYMOUS

"It is better to shoot for the stars and land in the trees, than to shoot for the trees and land in the mud." — ANONYMOUS

"Only in the dictionary does success come before work." — ANONYMOUS

"Winning is overcoming that person inside of us who wants to quit." — ANONYMOUS

"Olympic success is always preceded by Olympic motivation." — ANONYMOUS

"Not what men do worthily, but what they do successfully, is what history makes haste to record." — H.W. BEECHER

"Be a winner, not a whiner!" — ANONYMOUS

"That which we persist in doing becomes easier to do; not that the nature of the thing has changed, but that our ability to do it has increased."
EDWIN MARKHAM

"If we resist change, we'll fail, if we accept change, we'll survive; and if we create change, we'll succeed." – ANONYMOUS

"In the history of passions, each human heart is a world in itself; its experience can profit no others." – ANONYMOUS

"If you always do what you've always done, you will always get what you've always got!" – ANONYMOUS

"The best way to predict our future is to create it." – ANONYMOUS

"Trying to find happiness from the 'outside in' is a futile task. The only true happiness comes from the 'inside out.'" – ANONYMOUS

"Change is constant, it's everywhere. The only change I don't like is changing my one-year-old daughter's diaper." – ANONYMOUS